THE WAY OF THE CRUCIBLE

The Way

of the Crucible

Real Alchemy for Real Alchemists

Robert Allen Bartlett

Foreword by
Dennis William Hauck

Ibis Press
Lake Worth, FL

Published in 2009 by Ibis Press
An imprint of Nicolas-Hays, Inc.
P. O. Box 540206
Lake Worth, FL 33454-0206
www.ibispress.net

Distributed to the trade by
Red Wheel/Weiser, LLC
65 Parker St. • Ste. 7
Newburyport, MA 01950
www.redwheelweiser.com

First Edition 2008 by Lulu.com
This Edition 2009 by Ibis Press

Library of Congress Cataloging-in-Publication Data

Bartlett, Robert Allen.
  The way of the crucible : real alchemy for real alchemists / Robert Allen
Bartlett ; foreword by Dennis William Hauck.
      p. cm.
  Includes bibliographical references (p.   ).
  ISBN 978-0-89254-154-6 (alk. paper)
  1. Alchemy. I. Title.
  QD26.B3374 2009
  540.1'12--dc22   2009035785

ISBN 978-0-89254-154-6

Cover painting by Benjamin Vierling
www.bvierling.com
Cover design by Studio 31
www.studio31.com

Printed in the United States of America

CONTENTS

Kids! Don't try this at home!

The practice of Real Alchemy is inherently dangerous. Formal laboratory training is encouraged. Consulting a licensed physician is encouraged before consuming herbal preparations. Familiarize yourself with the laws that may apply to you in your jurisdiction and act accordingly.

Read as many of the other books on the subject of Alchemy as possible. (A list of recommended books can be found in the Bibliography.) Learn as much as you can from a qualified teacher. And above all, know the theory before attempting the practice.

This book is sold for informational purposes. The author and publisher will not be held accountable for the use or misuse of the information in this book.

FOREWARD

by Dennis William Hauck

In my work in alchemy over the last forty years, I have met a lot of people who call themselves "alchemists" who never set foot in a laboratory. In fact, just about anyone who talks about transformation—whether it is on the personal level, in the arts, or in business—is likely to call themselves alchemists. In the popular mind, the psychological and spiritual aspects of alchemy have overshadowed the work in the lab.

Traditionally, however, the alchemists were much more than intellectuals or philosophers. To accomplish lasting transformations, they had to succeed not only on the mental and spiritual levels but on the physical level as well. Tapping into the synergistic relationship between mind, matter, and energy is the essence of traditional alchemy and what differentiates it from all other disciplines.

How did the split between spiritual and practical alchemy come about? During the heyday of alchemy in the Middle Ages, alchemists around the world were pursuing the fabled Philosopher's Stone, which was said to instantly perfect any substance and change lead into gold. But the fascination with gold brought a new class of mercenary alchemists known as "puffers", who sat at their furnaces constantly fanning their bellows, hoping to produce gold by purely physical means.

The puffers worked only with external fires, never kindling the "secret fire" of the initiated alchemist that originated in psychological purification and spiritual meditation. When the puffers were unable perform their transformations, they resorted to trickery to finance their endeavours. Before long, a backlash developed against alchemists in which true alchemists suffered along with the puffers.

Gradually, alchemy split into two different directions. The purely physical work of the puffers, who discovered many new compounds and laboratory techniques, gave rise to modern chemistry. True alchemists rejected this commercialization of alchemy and were forced to practice their art in seclusion. In

1

their view, the central work in alchemy—the operation of the Stone—was not within reach of the chemist.

Today, however, the two paths of alchemy are converging again. Advances in quantum physics have revealed the hidden role consciousness plays in nature, and many other fundamental alchemical principles are being proven in practical demonstrations. The operations of alchemy have been shown to work in psychology, sociology, business management, and other seemingly unrelated areas.

The new alchemist is a hybrid born out of the two cultures of science and mysticism. This modern breed of scientist-mystic or chemist-alchemist is personified by the author of this book, Robert Bartlett. Robert does traditional alchemy using modern tools. Educated as a chemist, he was initiated into alchemy by Dr. Albert Reidel (Frater Albertus) and became his chief chemist at Paralab.

In his lab work, Robert integrates chemical knowledge and modern laboratory techniques with ancient alchemical wisdom. He also knows how to enhance his experiments with work on the personal and spiritual levels, and the first thing you notice when you meet him in person is his lack of ego and complete dedication to alchemy.

In short, Robert is the modern epitome of the true alchemist, and I highly recommend his work to both beginning and advanced students of the Art. This book and his previous introductory text, *Real Alchemy*, are among the few works available today that teach alchemy the way it was meant to be taught.

Dennis William Hauck is an author, consultant, and lecturer working to facilitate personal and institutional transformation through the application of the ancient principles of alchemy. As one of the world's few practicing alchemists, he writes and lectures on the universal principles of physical, psychological, and spiritual perfection to a wide variety of audiences that range from scientists and business leaders to religious and New Age groups. Hauck's interest in alchemy began while he was still in graduate school at the University of Vienna, and he has since translated a number of important alchemy manuscripts dating back to the thirteenth century. He is the publisher of the *Alchemy Journal* and on the board of directors of the International Alchemy Guild. His bestselling book, *The Emerald Tablet:*

Alchemy for Personal Transformation (Penguin Putnam 1999), presents startling new revelations about the mysterious "time capsule of wisdom" that inspired over 3,500 years of alchemy. His next book, *The Sorcerer's Stone: A Beginner's Guide to Alchemy* (Citadel Press, 2004), is an entertaining introduction to both practical and spiritual alchemy. His latest book, *The Complete Idiot's Guide to Alchemy*, is a modern apprenticeship for anyone interested in the ancient craft of alchemy. Hauck has been interviewed on nearly four hundred radio and TV programs, including such popular national shows as "NPR's Morning Edition", "Sally Jessy Raphael", "Geraldo", "A&E Unexplained", "Sightings", "Extra", "Martha Stewart Live", "The O'Reilly Factor", and "CNN Reports". He also serves as a special consultant to a several leading film production companies and a number of popular television programs. Website: www.AlchemyLab.com.

PREFACE

Many people today believe alchemy is just a psychological metaphor concerning the reintegration of our fractured psyche with no basis in the physical world. But any alchemist worth his salt would beg to differ. Alchemy is the ancient sacred science concerned with the mysteries of life and consciousness, not just in our own psyche but reflected throughout all of Nature.

I've been exploring alchemy for most of my life and especially its practical applications here in the world around us. The ancient sages advise us that understanding the workings of Nature in this world will aid in our understanding of more subtle realms from which Nature derives.

Perhaps I should start at the beginning and tell you how this book came about.

As a young boy, I had a fascination with rocks and minerals. I had collected an extensive array of interesting specimens and was constantly searching for more. It didn't take long to find out that there were things you could do to rocks in order for them to reveal their names. Scratch them on a tile plate, treat them with acid, mix them with common salts and heat them with a candle flame directed by a blowpipe; this was all great fun to me.

I scoured the local library for information on other experiments with minerals and quickly discovered the world of the alchemists. The artwork and mystic symbols struck a deep chord inside of me, and little did I know it would blossom into a guiding force that would follow me the rest of my days.

Very soon, I had commandeered my older brother's chemistry set and established a small laboratory in the corner of my bedroom. I studied all the texts about alchemy that I could find, but everyone seemed to be reticent and secretive about practical works.

I filled the house with dense white smoke a time or two, and blew a hole in the ceiling (honest, Dad, it was an accident), which prompted my parents to have me relocate my growing laboratory into a wooden shed we had in the backyard. That was perfect; now I had four times as much room! By the age of fourteen, I had constructed a gas fired furnace and tried my hand at distilling the mineral acids from copper sulfate and potassium nitrate, the vitriol and niter of the ancient sages. Imagine my joy when I successfully smelted lead from a lump of galena and later copper from a piece of malachite. Still, this was just chemistry and the secrets of the alchemist continued to elude me.

I began to study the writings of witches, wizards and astrologers to gain an understanding of the secretive language of the alchemists and became quite good at casting and interpreting horoscopes, brewing potions and casting a spell or two. My studies lead me into herbology and the mystical Qabalah as well as the higher magics of the Middle Ages.

By the end of high school, I was fully enmeshed in the teachings of The Golden Dawn as presented by Israel Regardie; in fact I devoured all of Regardie's writings. In one of his works, *The Philosopher's Stone*, I gained new insights into the world of alchemy, but most importantly a name, Frater Albertus of the Paracelsus Research Society in Salt Lake City, Utah. I was in college now, ostensibly earning a degree in chemistry but not learning what I was really interested in.

After purchasing *The Alchemist's Handbook* by Frater Albertus, I realized that here was a man teaching the very things I had been searching after. I quickly applied for admission to the classes offered at the Paracelsus Research Society, which later became Paracelsus College, recognized by the State of Utah.

In April of 1974, I attended the first of seven 2-week intensive training courses in the Hermetic Arts taught by Frater Albertus, one of the most well-known practical alchemists of the 20th century. My life changed entirely; I dropped out of college, moved to the primitive heart of Idaho, and devoted my time to the study and practice of alchemical works. I even worked underground in an antimony mine, an important resource for the practical alchemist.

In the third year of classes, Frater Albertus announced that Paracelsus Laboratories (Paralab) was being formulated and would soon be opening its doors. Paralab was to begin preparing herbal and mineral materials in the alchemical tradition and offer them for research and use in the alternative healthcare fields that were beginning to emerge at that time. At the end of class, we were allowed some private one on one talks with Frater Albertus. I expressed a burning desire to work at Paralab, and Frater Albertus encouraged me to return to school and finish my degree in chemistry; my work at Paralab would be waiting.

By 1979 I had received a bachelor's degree in chemistry and immediately moved to Utah.

In June of that year I was hired as the Chief Chemist at Paralab and for the next several years worked closely with Frater Albertus to develop a line of products derived from the mineral world.

Paralab proved to be a little ahead of its time and by 1983 was all but officially closed. With the passing of Frater Albertus in 1984, both Paralab and Paracelsus College closed their doors and the students seemed to be dispersed to the four winds.

I disappeared from the area for many years and pursued alchemical research privately. Being employed as a chemist allowed me to have access to state of the art chemical instrumentation by which I could follow alchemical experiments in ways undreamed of by the ancient masters of the Art. Over the years I have been able to amass a wealth of analytical data on a wide range of alchemical products, a task Frater Albertus charged me with many years ago and something I had always planned on publishing at a future date.

In 2003 I received a request to give a short presentation on practical alchemy at a local Healing Center here in the Pacific Northwest. The time scheduled was for a three-hour talk; however, it stretched into five hours and at the end everyone excitedly requested a second class so we could continue. Eventually it turned into three six-hour classes that have been well received in various locations around the area for the past five years.

During that time I received so many requests from students for a text they could follow along with that I decided to write *Real Alchemy* as a primer and introduction to the much larger work I had planned to publish all along; this is that book.

THE WAY OF THE CRUCIBLE

Herein you will find a wealth of information concerning the practical art and science of alchemy that makes this arcane subject accessible to the modern student.

INTRODUCTION

Our first book, *Real Alchemy*, provides a glimpse into the general theory and practices surrounding the Art of Alchemy as handed down in the Western World. With that groundwork laid, this second work delves more deeply into "the why of the how" behind laboratory alchemy and elaborates in greater detail on some of the mineral works of Alchemy both ancient and modern. Modern Alchemy? Yes, Alchemy has been an evolving, practical art for over two thousand years as an exploration of Nature, Reality, and Perfection of Self.

Today, there are many students, practitioners and resources available on the Hermetic Arts worldwide and the number is growing.

Alchemy has been called the "Perennial Philosophy" because all throughout history it has seen many revivals, followed by periods when it was subdued, even outlawed in some countries.

The practice of alchemy requires the participation of many different levels of reality. These constitute the many paths of the path, active at the physical, mental, and spiritual levels. They must all work together in unison because they are extensions of each other.

The language, symbolism and operative principles of alchemy rely on what is known as "The Law of Correspondences". The macrocosm and microcosm are intimately linked and inseparable. As the Emerald Tablet of Hermes says:

> That which is above is like unto that which is below, and that which is below is like unto that which is above.

In alchemical texts, one thing may be called by many different names, or many different things are called by the same name.

A key to much of this confusion can be found in the language of astrology, the sister science of alchemy. In times past, it was necessary to conceal one's involvement in the hermetic arts by the use of a code. The indelible symbols of the heavens provided just such a ready made code.

Astrology teaches that everything on earth has its correspondence in the sky, and everything in the sky has its correspondence on the earth.

A thing on earth is said to be ruled by a planet or sign in the heavens. This planet or sign rules other things on the earth having the same vibratory quality. The substance, principle, or quality designated by an alchemist may be called by the name of any one of the various objects ruled by the same astrological influence. Thus did the alchemists write and think in the language of celestial correspondences. Things on one plane correspond to definite things on another plane, which may be determined through their astrological rulership. Everything in existence on any plane vibrates in its inner nature to some astrological tone.

Paracelsus stated this principle of correspondence very concisely in his work titled *Paragranum*.

> If I have manna in my constitution, I can attract manna from heaven. Melissa is not only in the garden, but in the air and in heaven. Saturn is not only in the sky, but also deep in the earth and in the ocean. What is Venus but the Artemisia that grows in your garden? What is iron but Mars? That is to say Venus and Artemisia are both the products of the same essence, and Mars and iron are both manifestations of the same cause. What is the human body but a constellation of the same powers that formed the stars in the sky? He who knows what iron is, knows the attributes of Mars. He who knows Mars, knows the qualities of iron. What would become of your heart if there were no Sun in the universe? What would be the use of your Vasa Spermatica if there were no Venus? To grasp the invisible elements, to attract them by

their material correspondences, to control, purify and transform then by the living power of the spirit—This is true alchemy.

Alchemy is not just a psychological process, nor is it a simple spiritual allegory, or an outdated chemical demonstration. It is a harmonious blending of physical and subtle forces, linked together through hidden correspondences, which lifts the subject, be it man or metal, to a more evolved state of being.

This is Nature's intention all along; the artist follows nature's lead and seeks to assist her works and accelerate the process.

Alchemy

What is alchemy and where did it all start? The origins of alchemy are presently lost to us in the mists of prehistory; perhaps they will be rediscovered in some future time. The traditions have been passed down to us through several thousand years of practitioners and there are many theories as to when and where it all began. A gift of the Gods, a gift from angelic beings, remnants from an ancient advanced culture, a spiritual knowledge passed down by numerous patriarchs of several religions, the list goes on. Where and when are of lesser importance to us compared to the message itself. The alchemist Eireneaus Philalethes, writing in the late 1600s, put the whole matter to rest very simply, saying:

> If it is founded on the eternal verities of Nature, why need I trouble my head with the problem whether this or that antediluvian personage had a knowledge of it? Enough for me to know that it is now true and possible, that it has been exercised by the initiated for many centuries, and under the most distant latitudes; it may also be observed that though most of these write in an obscure, figurative, allegorical, and altogether perplexing style, and though some of them have actually mixed falsehood with truth, in order to confound the ignorant, yet they, though existing in many

11

series of ages, differing in tongue and nation, have not diversely handled one operation, but do all exhibit a most marvelous and striking agreement in regard to the main features of their teaching—an agreement which is absolutely inexplicable, except on the supposition that our Art is something more than a mere labyrinth of perplexing words.

Philalethes, *Metamorphosis of Metals*

Franz Hartmann, the German occultist and author of numerous works in the latter half of the 1800s, described alchemy in the following words:

Alchemy is a Science of Soul that results from an understanding of God, Nature, and Man. A perfect knowledge of any one of them cannot be obtained without the knowledge of the other two, for these three are one and inseparable. Alchemy is not merely an intellectual but a spiritual science, because that which belongs to the spirit can only be spiritually known. Nevertheless, it is also a science dealing with material things, for spirit and matter are only two opposite manifestations or poles of the eternal One.

Alchemy in its more material aspect teaches how minerals, metals, plants, animals, and men may be generated or made to grow from their "seeds". In other words, how that generation, which is accomplished during long periods of time in the due course of the action of evolution and natural law, may be accomplished in a comparatively short time, if these natural laws are guided and supplied with the proper material by the spiritual knowledge of man.

Alchemy is also an art, and as every art requires an artist to exercise it, likewise this divine science and art can be practiced only by those who are in possession of the divine power necessary for that purpose. It is true that the external manipulations required for the production of certain alchemical preparations may, like an ordinary chemical process, be taught to anybody capable of reasoning. However, the results that such a person would accomplish would be without life, for only he in whom the true life has awakened can awaken it from its sleep in matter and cause visible forms to grow from the primordial Chaos of nature.

Alchemy in its highest aspect deals with the spiritual regeneration of man and teaches how a god may be made out of a human being or, to express it more correctly, how to establish the conditions necessary for the development of divine powers in man.

From *Alchemy* by Franz Hartmann

Throughout much of its long history, alchemy has been described as the quest for the Philosopher's Stone, that mystical agent of transformation; "the medicine of men and of metals". For as the alchemists agree, the intention of Nature is that all metals attain to the state of gold, the perfect and incorruptible metal. Through the Law of Correspondences, this attainment is true for all three kingdoms and planes of existence.

Alchemy is the art of perfecting the metals, say the ancient sages; not only the metals of the forge, but those same metallic essences as reflected in our constitution. We see a shadowy reflection of this connection in our daily speech, such as having a heart of gold, or nerves of steel, or a cast iron stomach. Each of our experiences in life can be attributed to the metals, from the leaden heaviness of grief, sorrow and poverty to the power and prestige of gold. Our attitude to the experience will determine whether it shines with metallic value or becomes tarnished by corruption.

The Philosopher's Stone is said to heal metals of their diseases, which are the source of their corruptibility, and remove the impurities that hinder their perfection into gold.

In addition to the "Stone of the Wise" is the preparation of the "Elixir of Life", said to confer a long and healthy life to the one who partakes of it; health not only of the body but of the mind and spirit. After obtaining our gold, we would certainly want to live long enough to enjoy it to the fullest.

The goal of the alchemist is the search for perfection, bringing health on all levels, body, soul, and spirit; complete freedom from disease coupled with longevity and spiritual clarity. For man or metals, this is the real transmutation, this is the Great Work of the alchemist.

The alchemists admonish us to follow Nature, for Nature generously provides for all her creations. However, Nature has all of time to bring things to their destined state of perfection. The alchemical art is concerned with helping this transformation along more quickly, whether it be the perfection of metals or of ourselves.

Frater Albertus defined alchemy very succinctly as "consciously assisted evolution" and "raising the vibrations". It is simply following Nature's own laws, but we must come to an understanding of what these laws are before we can be of assistance.

The work of the alchemist is concerned with the preservation of vitality and consciousness within his subject and ultimately provoking its evolution into a more refined being.

The changes we see in the laboratory do not take place without a corresponding change of conditions on other planes of existence.

The crude body of matter is a reflection of its disorganized and immature spiritual counterpart. The processes of alchemy seek to remove the hinderances to this spiritual maturity, leading to the formation of an incorruptible spiritual body and its reflection in the outer world.

External Influences

Nature has a rhythm—there are cycles within cycles occurring all around us at all times, like a symphony; everything is in motion, everything vibrates. The seeming differences between things are entirely due to varying rates and modes of vibration. The alchemist seeks to follow Nature and keep in time with the music, the constant flow of consciousness.

In the laboratory, we can control many of Nature's processes at our whim, but there are certain external influences we have no control over which are also important to the alchemical art. The Earth radiates a living field of energy and is constantly rained down upon with the energies of the cosmos.

The influence of the Sun and Moon are especially important to the alchemist; they are the outer garment of the spiritual fire, the Celestial Fire, we seek to incorporate into our matter, or we seek to awaken the inherent fire already residing within.

As a practical example, during the processes involving the deliquescence of salts, this is not just a simple dissolution of the matter; instead we seek to infuse into our subject the life energy radiating from earth, and the celestial fire raining down from the heavens. The medium of exchange for this is the moist vapor which surrounds and permeates our planet. Water is the carrier of these subtle energies. This vapor is the exhalation of the earth, or "Spiritus Mundi", the Cosmic Breath.

Our matter draws in this energy, concentrating and condensing it into a vehicle we can work with. The ancients called a properly prepared salt a "magnet", which becomes the vehicle for the vital essence infused into it by the process of deliquescence.

Although our deliquescence process is of a relatively short duration, we have the option of selecting the timing of celestial events and "catching a favorable wind". Think about plants, getting infused all year to develop their healing potentials and then think about minerals getting infused for millions of years, and you will understand why the sages have all claimed that the most powerful medicines are to be found in the mineral realm.

Cycles of the seasonal energies/elemental influence

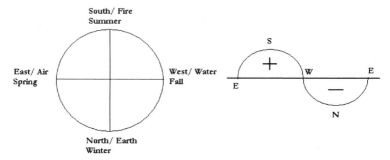

In laboratory works, we try to keep within Nature's cyclic flow of energy by timing operations compatible to the season.

The cycle of the seasons, the pulse of Nature, is the first and most obvious of the external influences which affect alchemical works. The practical alchemist can learn a great deal by working in a garden and observing the course of Nature throughout the year.

The positive phase of the year, from the spring equinox to the autumn equinox, is dominated by the Volatile, active elements of Air and Fire. During this time the vital essence and Celestial Fire are actively on the rise. This is the time to begin new works. The deliquescence of salts to capture Celestial Fire, processes of exaltation such as distillation and sublimation, regeneration by cohobation and circulations are all favored at this time of the year.

The negative phase of the year, running from the autumn equinox through winter to the spring equinox, is dominated by the Fixed energies of Water and Earth. The life force begins to hibernate and things start to fall apart. This is a time for works of fermentation, putrefaction, incineration and calcination. A time to separate the pure from the impure. It is also a time to work on salts by dissolution and recrystallization.

Each day of the week is under the rule of a particular planetary energy; in fact our names for the days are derived from the planets themselves or from the names of gods and goddesses associated with the planet. These associations are shown in the table below.

PLANET	WEEKDAY
Sun	Sunday
Moon	Monday
Mars	Tuesday
Mercury	Wednesday
Jupiter	Thursday
Venus	Friday
Saturn	Saturday

Laboratory works can be favorably enhanced by keeping within the timing of a compatible planetary energy flow. For example, if we are working with a Venus ruled herb, we should schedule the various operations to be performed on Fridays. In a more generalized way, we might take advantage of the Moon's influence over liquids to perform dissolutions or the fiery nature of Mars to calcine a material, or the cold, limiting qualities of Saturn to crystallize a salt.

The cyclic manifestation of planetary energies are also broken down into "hourly" effects, and even into minutes and seconds of "planetary influence". However, as a practical limit, consideration of planetary hours is as far as we need to go.

The concept of planetary hours derives from the ancient Chaldean astrologers who divided the time from sunrise to sunset into twelve equal segments called the day hours, and from sunset to sunrise into twelve segments called the night hours. The progression of these hours is always the same, running from Saturn to the Moon in the same order as the spheres arise on the Qaballistic Tree of Life. Thus we have the following order: Saturn, Jupiter, Mars, Sun, Venus, Mercury, Moon, then back to Saturn again and the cycle repeats. The first hour after sunrise is always ruled by the same planet that rules that particular day, and thus it is an especially powerful time for working with that planet's energies.

The traditional method for calculating the planetary hours involves finding the times of sunrise and sunset for a specific location and dividing that interval by twelve to obtain the length of the planetary hour for that day. The length of these hours changes by about four minutes each day, being longer in the summer and shorter in the winter. As mentioned, the first

17

hour of the day belongs to the planet which governs the day, and then the succeeding hours follow the order given above.

The same process is applied to the time interval between sunset and sunrise, to obtain the night hours. It all seems like a nightmare of calculation, but there are tables and programs you can find on the Internet which do all the work for you; all you have to know is what day it is and where in the world you are.

There is an easier method which is popular among many artists and was the method favored by Frater Albertus at the Paracelsus College. The method derives from an earlier Rosicrucian teaching.

In this method the twenty-four-hour day, from midnight to midnight, is divided into seven equal periods. The periods follow the same order as above, and the time period that encompasses sunrise (3:26 to 6:51 am) is ruled by the planetary ruler of that day. The following table illustrates the divisions which are used year round and based on local time for any location.

Planetary Hours chart

Period	Sun	Mon	Tues	Wed	Thur	Fri	Sat
12:00 to 3:25 am	Mars	Mercury	Jupiter	Venus	Saturn	Sun	Moon
3:26 am to 6:51 am	Sun	Moon	Mars	Mercury	Jupiter	Venus	Saturn
6:52 am to 10:17 am	Venus	Saturn	Sun	Moon	Mars	Mercury	Jupiter
10:18 am to 1:42 pm	Mercury	Jupiter	Venus	Saturn	Sun	Moon	Mars
1:43 pm to 5:08 pm	Moon	Mars	Mercury	Jupiter	Venus	Saturn	Sun
5:09 pm to 8:34 pm	Saturn	Sun	Moon	Mars	Mercury	Jupiter	Venus
8:35 pm to 12:00	Jupiter	Venus	Saturn	Sun	Moon	Mars	Mercury

Now, the election of a suitable time to operate is a bit easier. Let's say we wanted to start an extraction of a Mercury ruled herb. Wednesday, between 3:26 am and 6:51 am, would be ideal, but we have other appointments to keep and cannot do the work until the weekend. Saturday has a Mercury hour between 8:35 pm and midnight, but the influence of Saturn is constrictive and cold, not very conducive to our present aim. Sunday has a Mercury hour between 10:18 am and 1:42 pm and the exalting influence of the Sun will add power to the extraction process, so this would be a much better time to do our work.

For all of these elections of time, get to know the nature of the planets so you will be able to determine if their influence will help or hinder the work you have planned.

Since all of the various systems of planetary hour division agree that the hour of sunrise is ruled by the planet of that day, it is always your safest bet for important works.

Moon cycles/ Lunar Mansions

Most of the ancient cultures based their calendar on the cycles of the moon. The effect of the moon on the ocean's tides is well known. Did you know this tidal influence also affects the atmosphere and even the earth's crust? In fact everything on our planet is heavily influenced by the course of the moon.

Some believe that the earth and moon share the same subtle etheric body and the cycles of the moon modulate this energy body in a predictable manner.

The practical alchemist utilizes this modulated energy to assist works in the laboratory. The first and most obvious lunar cycle is that of the waxing and waning of the moon each month. The Waxing Moon is good for Enriching an Essential element by circulations, sublimations, or distillations. Its magnetic influence draws things up, volatilizing, exalting, spiritualizing them.

The Waning Moon is good for Separating the Pure from the Impure whether by fermentation, extraction, or calcination, etc. Just as the dying moonlight, our matter is subjected to the fermentation and putrefaction of death in order to release its essence, thus separating the pure and the impure.

Each month, as the moon circles around us, the signs of the zodiac become a backdrop to it, one by one. The moon concentrates these zodiacal energies and focuses them upon the earth. In this way, the alchemist has an opportunity to harness the power of a particular sign in his operations on a monthly time scale. The traditional correspondence between signs and operations is shown below.

Sign	Operation	Sign	Operation
Aries	Calcination	Libra	Sublimation
Taurus	Coagulation	Scorpio	Separation
Gemini	Fixation	Saggitarius	Incineration
Cancer	Dissolution	Capricorn	Fermentation
Leo	Digestion	Aquarius	Multiplication
Virgo	Distillation	Pisces	Projection

Another lunar cycle involves its daily motion of about 12 degrees and 51 minutes, through the zodiac. This gives rise to what are known as the 28 Mansions of the Moon. Each day, the moon resides in a different "Mansion", and each mansion has a specific influence. The table below lists the 28 Mansions by their name, derived from the names of stars closest to their starting point. Also listed are the sign and planetary ruler which indicate the type of influence imparted by the moon.

Mansions of the Moon

#	Name	Sign	Start Deg:Min	Planetary Ruler
1	Alnath	Aries	00:00	Sun
2	Albotain	Aries	12:51	Moon
3	Azoraya	Aries	25:43	Mars
4	Aldebaran	Taurus	08:34	Mercury
5	Almices	Taurus	21:26	Jupiter
6	Athaya	Gemini	04:17	Venus
7	Aldirah	Gemini	17:09	Saturn
8	Annathra	Cancer	00:00	Sun
9	Atarf	Cancer	12:51	Moon
10	Algebha	Cancer	25:43	Mars
11	Azobra	Leo	08:34	Mercury

12	Acarfa	Leo	21:26	Jupiter
13	Alahue	Virgo	04:17	Venus
14	Azimech	Virgo	17:09	Saturn
15	Argafra	Libra	00:00	Sun
16	Azubene	Libra	12:51	Moon
17	Alichil	Libra	25:43	Mars
18	Alcalb	Scorpio	08:34	Mercury
19	Exaula	Scorpio	21:26	Jupiter
20	Nahaym	Saggitar	04:17	Venus
21	Elbelda	Saggitar	17:09	Saturn
22	Caadaldeba	Caprico	00:00	Sun
23	Caadebolach	Caprico	12:51	Moon
24	Caadacohot	Caprico	25:43	Mars
25	Caadalhacbia	Aquar	08:34	Mercury
26	Almiquedam	Aquar	21:26	Jupiter
27	Algarfalmuehar	Pisces	04:17	Venus
28	Arrexhe	Pisces	17:09	Saturn

The system of Lunar Mansions used in the West was developed by the ancient Chaldeans. This system became very popular with astrologers and magicians of the Renaissance period, for election of times to undertake various actions. In laboratory alchemy, it provides another guide for opportune times in which to perform operations and capture the momentum of celestial energies.

In the use of astrology to prepare medicines specific for an individual, the advice Nicholas Culpepper gives in his *Complete Herbal*, written in 1652, is still of great utility:

> To such as study astrology (who are the only men I know that are fit to study physick, physick without astrology, being like a lamp without oil) you are the men I exceedingly respect, and such documents as my brain can give you at present (being absent from my study) I shall give you, and an example to shew the proof of them.
> 1. Fortify the body with herbs of the nature of the Lord of the Ascendent, 'tis no matter whether he be a Fortune or Infortune in this case.

21

2. Let your Medicine be something antipathetical to the Lord of the Sixth.
3. Let your Medicine be something of the nature of the sign ascending.
4. If the Lord of the Tenth be strong, make use of his Medicines.
5. If this cannot well be, make use of the Light of Time.
6. Be sure always to fortify the grieved part of the body by sympathetical remedies.
7. Regard the Heart, keep that upon the wheels because the Sun is the Fountain of Life, and therefore those Universal Remedies, Aurum Potable and the Philosopher's Stone, cure all diseases by fortifying the heart.

Inner and Outer Work

Laboratory work as meditation or spiritual exercise develops concentration and focus as well as a little dexterity when performed with intention, as all alchemical works should be.

Mental and spiritual practices of the alchemists, working in conjunction with the products of the laboratory, work upon transforming all levels of the operator. You become fully involved, body, soul and spirit.

Before we go any further, you should remember that notebooks and note taking while studying alchemical texts are part of your personal alchemical fixation. Also it is important to make notes, and draw diagrams or flowcharts of the works you plan to attempt. Taking the time to write down your notes and thoughts fixes them firmly in the subconscious mind. They also serve as a logbook and documentation of your progress.

You can use a bound notebook or a loose-leaf binder to allow additions or rearrangement.

Later, during periods of Meditation, which is part of the alchemical circulation, your thoughts and concepts are digested and matured into a fruitful harvest of insight.

The nature of internal works is a personal choice of the individual. Alchemy is a compatible foundation for all: qabalist, pagan, Christian, Hindu, Taoist, or whatever religious or philosophical ideals you embrace.

One part of the internal work of the alchemist which we might call Mental Alchemy involves the transformation of our habitual thought patterns. Our thoughts have power to attract circumstances into our lives and these circumstances are related to the types of thoughts we constantly entertain. If we are filled with anxiety and fear, we attract events that are of a similar vibration. If we are positive, determined, and optimistic, we attract a corresponding type of circumstance into our lives.

Each type of thought has an astrological association to a particular planet and thus, through the law of correspondence, represents the metals upon which we work to transmute.

Just as the various types of tissues of the physical body, such as bone, nerve, and muscle, are organized into the structures that make up the body, the subtle substance of thoughts are organized into structures within our subtle body according to their specific range of vibration; these are our interior stars.

Chart of the classical virtues and vices associated with each planet

Planet	Vices	Virtues
Saturn	Greed	Charity
Jupiter	Cunning	Wisdom
Mars	Anger	Courage
Sun	Pride	Humility
Venus	Lewdness	Virtue
Mercury	Envy	Benevolence
Moon	Lust	Stability of Mind

Upon a yet higher level come the activities we might term Spiritual Alchemy. All of the circumstances and events handed to us by nature are the crude metallic ores upon which we must work. Here we learn to separate the dross of external appearances from the true metal, which is the central lesson we need to acquire for continued growth. It is our attitude towards the events and our power of discrimination which separates the pure grain of metal from the impure matrix. Each of these grains of precious metal is a most subtle matter which enters into the formation of an incorruptible spiritual body. Each type of lesson provides a subtle tissue based on its astrological affinity, so we seek to acquire the lessons from each of the seven ancient metals of alchemy.

The active agent affecting all of these transformations is fire. In the laboratory, we use fire to heat a substance, which in turn raises its vibratory rate. On the mental level this fire is emotion, the fire of desire and enthusiasm, the intensity of pleasure and pain. At the spiritual level it is the fire of aspiration and inspiration.

Although it is convenient to classify them as material, mental and spiritual alchemy, they are all really one Art, the alchemy of transformation. As we proceed through the various sections ahead, we will develop these ideas in greater detail.

CHAPTER ONE

East and West—Medicines vs Gold

The book *Real Alchemy* examines the Western alchemical tradition with ideas and methods followed by ancient as well as contemporary artists. The ancient cultures of China and India also developed alchemy into a high art. The terms may be different from those of the West but the philosophical principles are the same.

Chinese laboratory alchemy employs many exotic materials like jade, pearls, mercury, and arsenic, in medicines and in the quest for the Elixir of Immortality.

Indian alchemy has always been about preparing superior medicines that can work on the physical and subtle parts of man's constitution.

The transmutation of base metals into gold was also part of Chinese and Indian alchemy, but never considered the primary goal of the art.

The West is always going for the gold, and for a long time the whole focus on alchemy was making gold; they became synonymous in the popular mind. The art of alchemy became very distorted, labeled fraud, even outlawed during various periods of history.

Paracelsus (1493-1541) changed that idea around and directed the focus back onto superior medicines prepared through the alchemical art, and lifting mankind from a dark age. His Memorial plaque reads:

> Here is buried Phillippus Theophrastus Paracelsus, the distinguished doctor of medicine, who *by wonderful art* healed malignant wounds, leprosy, gout, dropsy and other incurable diseases of the body, who gave his possessions for distribution among the poor.

We will be examining alchemy in both Western and Eastern terms throughout this book. Eastern alchemical ideas represent a tradition which is over 5000 years old, so it is useful for the practicing alchemist to be familiar with this body of knowledge.

The writings from the East very often shed light on subjects which were only disclosed under a heavy veil of secrecy and symbol in the West. We can use them to help guide us through the labyrinth of alchemy.

Taoist alchemy, Ayurveda, and Alchemical Medicines

Even in very ancient times, India and China had communication and in both lands, alchemy developed along similar lines; the terms and associated mythology may have differed but most of the basic concepts are the same or at least closely parallel.

Taoist alchemy of China developed in two parts: *nei tan*, which is an internal process in which the body and physiological fluids of the alchemist himself formed the vessels and ingredients of the work; and *wai tan*, which involves laboratory alchemical works. These manual operations provide exterior materials as medicines which assist the process of the Great Work of creating an incorruptible body and elevating man's finer essence.

In India, there developed two medical traditions which shared the basic tenets, one to the north called Unani (which received Persian, Islamic, and Greek influences), and a more indigenous tradition to the south known as Siddha. Together they are collectively referred to here as Ayurveda, and comprise a system of internal processes coupled with very powerful medicines to assist one's evolution toward perfection.

Ayurveda literally means "The Science of Life". Indian alchemy, and particularly the Siddha tradition, developed a wide variety of alchemical processes for the preparation of metals used in medicines for regeneration of the body and transmutation of metals, in which metallic mercury played a key role. The texts concerning this part of alchemical works are known as Rasa Shastra, one of the eight branches of ayurveda and traditionally held to have been vouchsafed to mankind by the god Shiva himself. Much of the alchemical art as practiced in the West may in fact be derived from these ancient texts which made their way into the Arabian countries along the old spice trading routes and eventually found their way into the Western world, possibly even in the Library of Alexandria. The goals of Rasa Shastra are to establish perfect health of the body and mind, enhance longevity, and effect the transmutation of base metals into gold. Sound familiar?

In his book concerning the life and doctrines of Paracelsus, Franz Hartmann presents the possibility that Paracelsus traveled to India and studied there for nearly 15 years. Even a casual pursual of his works will reveal the remarkable similarities to the concepts put forward by the Indian adepts of the time.

The common thread running through all of these ancient traditions is that alchemy presents its adherents with a truly holistic approach to individually effecting the changes necessary in body and mind, which result in perfection of self and spiritual enlightenment. We are the base metal that becomes transmuted into incorruptible spiritual gold.

CHAPTER TWO

Ayurveda—Science of Life

The concept of Tridosha is of central importance to ayurveda and formed one of the important subjects taught by Frater Albertus at the Paracelsus College. Manfred Junius also mentions the close connection of ayurveda and Western alchemy in his excellent work *The Practical Handbook of Plant Alchemy*.

In this chapter, we present an overview of this Indian "Science of Life", which provides valuable keys to understanding our Western "Science of Life", called Alchemy.

In this system, as in Western alchemy, the ultimate source of All is consciousness, infinite living mind. The principle of polarity, giving rise to the universal essences or "Mahagunas", is expresssed as Purusha and Prakruti or Shiva and Shakti, the divine couple (Celestial Niter and Celestial Salt in the West).

The gunas are universal qualities through which we perceive Nature. This is similar to Aristotle's diagram of the elements coming forth from the qualities of hot, cold, wet and dry.

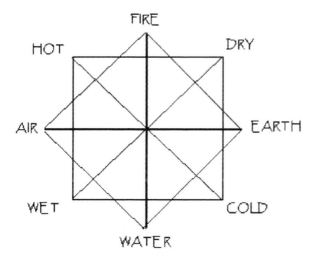

The Indian alchemists perceived ten pairs of opposites through which reality manifests.

29

The three alchemical essentials of Mercury, Sulfur, and Salt, in a most subtle state, that is, their essence, are called "Sattva, Rajas, and Tamas" (see chart 1 in appendix I); they represent the essential characteristics of consciousness. They ultimately bring forth the "Five Elements" (Earth, Water, Air, Fire, and Ether (Akasha or Quintessence) with their associated sense (odor, taste, touch, form, sound) and the interactions between them.

The elements in turn form the three essentials of the natural world, called Vata (Air), Pitta (Fire), and Kapha (Water), which express through a body. Don't let these terms confuse you; these are the same three essentials of Mercury, Sulfur, and Salt respectively, in the Western tradition. This blending of elements also gives rise to the six tastes which we will discuss later.

With some minor adjustments to our diagram of Western alchemical ideas, we can present the ayurvedic concepts of the Elements and Three Essentials as follows:

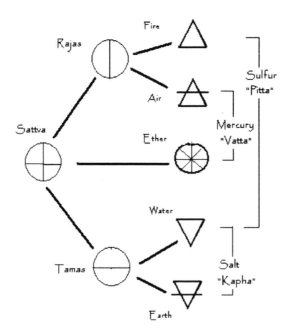

The three essentials here are called Doshas, and everything has a unique blending and balance of the three.

Each of the doshas (vata, pitta and kapha) is also subdivided into five types or subdoshas; thus there are five types of vata, and five types of pitta, and kapha, each with specific properties within its dosha. The qualities associated with each dosha are listed in chart 2 of appendix I.

When our particular blend of the three doshas becomes unbalanced, our bodies begin to malfunction leading to disease and premature aging. The word *dosha* actually translates to "defect", as the imbalance of the doshas is the cause of all the problems.

The key concept in ayurvedic medicine is that everything we eat, drink, think, and do has power to increase or decrease the influence of the doshas within ourselves and in anything else for that matter. In this way, the food we eat, the thoughts we entertain, and the types of activity we perform can be a medicine or a poison to us.

Ayurvedic physiology recognizes that each of the doshas has unique places of residence in the body and systems of circulation. The physical vehicles of these subtle energies form the structural components of the body, called Dhatus or tissues.

There are seven dhatus, listed as plasma, blood, muscle, fat, bone, marrow, and reproductive essence. These tissues are formed during the process of digestion and each has an associated waste product that must be removed from the body in order to maintain our dosha balance.

Each tissue nourishes another until the process of refinement and transformation of essences derived from food, drink and air produces the life-giving and sustaining "Ojas" that pervades the entire body.

Ojas is said to be the finest product of digestion, preventing disease, giving luster to the skin and regenerating the whole body; it is a part of man's quintessential nature. Ojas is part of a trio of sublimated qualities we will look at later.

So, when we eat normally, the food is first refined into a nutritive essence (plasma) with urine and feces as the waste product. This plasma is refined and incorporated into blood with mucus as the waste product. Blood is refined into flesh or

31

muscle with bile as the waste product. Muscle is further refined into fatty tissues with excretions from the eyes, nose, and throat as waste. Fat is refined and the essence moves into bone with sweat as waste. Bone is refined into marrow with hair and nails as waste, and marrow is refined into reproductive essence (sperm or ovum) with skin oils as waste. The reproductive essence is further refined into Ojas and this is the difficult part.

We naturally produce a certain amount of Ojas which maintains health, but it is hard to produce a sufficient amount which will allow the full expression of its power to perfect us body, soul, and spirit; this is where alchemical medicines come in.

Ojas itself is further refined into Soma, "The Nectar of the Gods", The Elixir of Immortality. Soma is the super refined essence that sustains Agni, the fire and light of consciousness.

In a sense the doshas are the waste products of the three Mahagunas (sattva, rajas, and tamas), which are the essential qualities of consciousness. The doshas provide a medium for the expression of a unique consciousness in the physical world. Our food and drink, the air we breathe, the actions we perform, as well as our mental/emotional habits all affect our elemental or doshic balance and thus how we reflect the three gunas.

Orchestrating this whole process of refinement is the digestive fire (Agni) which also has its several types forming the various hormones, enzymes, etc., involved with digestion and transformation of essence in each of the tissues. The concept of Agni is much more than just digestion of food and we will come back to it later.

The products of digestion, including waste, are circulated through the body by a series of ducts or channels (Nadis) which allow communication and transport between the tissues (Dhatus). There are twelve main duct systems (fourteen for women, including menstruation and lactation) physically within the body, and many hundreds of intangible channels which are collectively called "Srotamsi".

Ideally we eat foods and perform activities which keep our particular blend of doshas balanced and strengthened, our various channels unobstructed, our digestive fire strong and not overtaxed by excess or the wrong types of food for our nature; and this includes the type of lifestyle we have as well as our habitual mental/emotional states. Each of these things affects the balance of our blend of doshas, our constitution. The diet we have ultimately becomes digested into consciousness.

When our doshas become unbalanced, the digestive fire is thrown out of control, being either too hot or too cold to function correctly. This produces materials toxic to our systems such that disease conditions can arise.

These toxins can be divided into three main types. The first is called "Ama", which is a sticky waste product of incomplete digestion; it can build up in the digestive system. Ama is the most common toxin we accumulate through eating the wrong foods or overloading our digestion with excesses. Ama makes you feel heavy, weak, and tired.

If ama continues to build up over time, it can spill over into the various duct systems and begin circulating through the body. It then generally settles in some weakened organ or part of the body and obstructs the flow of vital fluids.

After ama stagnates here awhile, it becomes a more toxic and reactive substance described as "amavisha". Visha means irregular, unstable or poisonous.

Amavisha can react with each of the seven tissues (Dhatus) and their waste products as well as the sub-doshas of the system it is lodged in, to create the various disease symptoms. Depending on the mixture and location, these effects can be physical, emotional, and/or mental.

A third type of toxin we can accumulate is from external sources, from our environment, and is called "Garavisha" (artificial poisons). This class of toxin includes pesticides, preservatives, food additives, spoiled food, as well as heavy metals (like lead, cadmium, mercury, and arsenic); even household products like detergents and the quality of air we breathe can be sources of garavisha.

The alchemical work is concerned with the removal of all these toxins, establishing the balance of the doshas proper to our constitution, clearing out obstructions in the various channels, and the perfect refinement of substance and mind.

Body posture, breath, diet (physical and mental) and specialized medicines are all part of the raw materials whose regulation is necessary to bring us into a higher state of perfection. We cultivate consciousness.

CHAPTER THREE

The Art of Alchemy

A few years back, I was teaching an introductory class on the fundamentals of alchemy. After the class, a student approached me and asked, "How does one incorporate these ideas into everyday life?" The short answer was that the Hermetic Philosophy is a way of looking at life and like ayurveda it is the science of life. Life is the alchemical process going on around us everywhere and at all times. The alchemist strives always to live in harmony with Nature and to assist Her in bringing things to their highest state of perfection. And this applies most importantly to the work upon ourselves in order to achieve a happy, healthy, holy life as Nature intends.

In addition to our work in the laboratory, there are many practices we can adopt in our daily routine which can assist us on the path of the Great Work.

Food, water, and air form the prime materials our bodies need in order to function correctly.

The type of lifestyle we lead, including our habitual thought and emotional patterns, provides the subtle food that keeps our mind healthy. And both of these, substance and thought, interact with each other as well, both are aspects of the One.

Cooking with alchemy

"Let food be your medicine." —Hippocrates

A proper diet is crucial for self-transformation. Cooking is alchemy; we prepare our subject with a regulated heat and adjust its elemental balance with spices, then it becomes our elixir of life. The inherent qualities of different foods will increase or decrease our unique elemental or doshic balance, so we need to become aware of those qualities and how they affect the doshas.

Some general eating guidelines:
1. Try to eat organically grown foods as much as possible and prefer those that are fresh off the vine from local growers; they contain much more solar prana.
2. Clean all of your food carefully to remove external impurities as much as possible.
3. Prefer warm freshly cooked foods. Cooking is an art which assists the digestion and assimilation as well as the doshic balance of a meal.
4. Don't eat leftovers which are older than five hours; they've lost most of their vitality and are beginning to break down.
5. Use cookware which is made of earthenware, glass, iron or stainless steel.
6. Cook only when you are in the right frame of mind. Our thoughts and emotions influence the whole meal during preparation.
7. Avoid improper food combinations.
8. Prefer foods which are more Sattvic in nature. (More on this later.)

The manner in which food is cooked will also affect the doshas. Boiled foods are more vata balancing because they are hot and moist, which is opposite to vata's cold and dry qualities. Baked foods are more kapha balancing because they are hot and dry, the opposite of kapha's cold and moist character. Raw foods will increase vata with their cold, hard, rough qualities, and so they won't be digested well by someone with predominant vata qualities. Stir-frying with oil balances vata and if the oil is ghee, then pitta is also balanced. Steaming helps to balance pitta and kapha.

The lessons we learn in the kitchen will go a long way in guiding us in the laboratory.

The digestion process is our central fire and must be carefully tended. A weak digestion is kindled slowly to flame with the proper foods and spices instead of being smothered with improper fuels. What we eat also affects our thought and emotions because the digestive process includes how we digest our experiences. Food becomes consciousness.

Alchemy and the spice trade have a long association. The early "Spice Roads" connected China and India with "The Arabias" and provided exotic materials and information. Indeed the Arabian countries controlled the spice trade to the western world for centuries. We use spices to adjust the taste of food to our palate, though not always with good judgment.

Eating the proper diet for your particular elemental balance maintains health of the body and the mind. Learning the effects of various foods and spices on the doshas is the first step. Most of this is just common sense; for example, hot, spicy foods will increase the fire element or pitta. Cold and wet foods will soothe an active fire.

We are going to examine some of the factors involved in determining the qualities of a material and how a particular substance, food, spice, or whatever will affect our balance of elements.

This information forms the basis for why a particular medicine is applied and the art of preparing effective herbal combinations.

Count St Germain, an alchemist of the 17th century, was known to be very particular about preparation of his food when he was out and about. He is also known for his excellent health and longevity.

Doctrine of Signatures/Gunas/VPK factors/6 tastes

We mentioned the "Gunas" and the six tastes earlier as essential qualities in Nature.

These are the properties we experience through the senses as reality, and everything expresses itself through these properties. The combination of influences surrounding a particular material creates a sort of signature which we can read in order to understand its essential character. This has been called "The Doctrine of Signatures".

Chart 2 in appendix I lists these essential qualities and their relation to the doshas. The elemental quality of a thing is assessed by noting which qualities it manifests. For example, look at a lemon; its warm yellow color, oily peel, sharp and sour taste indicate a lot of pitta qualities, but it has a cooling nature and lots of juice, which are kapha qualities. The peel is somewhat bitter and that is a vata quality.

The rule of usage is that "like increases like" and "opposites balance". Hot spicy foods will increase the hot spicy qualities of pitta and balance the cold of vata and kapha; it's common sense. If you have a lot of pitta in your constitution, you might be prone to acid indigestion or skin rashes, or be irritable, even aggressively hostile; eating foods which increase pitta qualities is just going to make things worse. To regain balance you would choose foods and spices which are cooling and soothing.

If your constitution has an excess of vata qualities, you may experience various aches and pains, nervous disorders, worry, and anxiety. Cold and rough foods like raw vegetables will increase vata, as will dry foods like crackers and chips. To bring balance you would choose foods which are warm, soft and oily, and tastes which are sweet, salty and sour.

When we venture out into Nature, we can observe the various qualities at work around us and learn to read their signatures.

As a general rule, vata predominant plants have sparse foliage, very cracked bark, crooked and gnarled branches. They also have spindly growth habits and contain little sap due to vata's dry qualities. Pitta type plants have bright flowers, and moderate strength and sap content. Although they tend to grow in straight lines, they can also exhibit the spreading qualities of pitta, can be full of stickers and can have poisonous or burning qualities as well. Kapha plants show luxurious growth with abundant leaves and rounded lines. They have heavy, succulent leaves and stems containing a lot of water and sap.

The taste associated with each plant also provides valuable information on their elemental or planetary rulerships.

The Six Tastes and their Elements

Taste provides the body with important information concerning the digestion of the food it is receiving. The taste of a substance represents how it was formed from the basic elements of the universe, what therapeutic properties it has, and which organs it will affect. In preparing a meal, we can use mixtures of common spices to adjust the qualities it expresses, thus making it more compatible with our elemental balance.

Taste affects not only the body, but the mind and emotions as well. We use these connections in our everyday speech like describing someone as "sweet" or feeling "bitter" about some incident.

The Elements of Taste

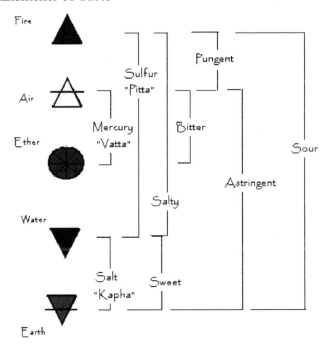

Sweet

The sweet taste is said to increase the vital essence of life. It is found in many foods, such as sugar, milk, rice, dates, bananas and raisins. It increases Kapha and decreases Vata and to a lesser extent, Pitta. It is a cooling taste, which helps in building all seven body tissues. With proper use, it increases strength and promotes longevity. In excess, it causes congestion and laziness, builds toxins and causes obesity. In moderation, it promotes stability and a sense of contentment, and is associated with the emotions of love, compassion, happiness, and bliss. It is made of the elements earth and water.

Sour

The sour taste is found in fermented foods and citrus fruit, sour cream, many types of cheese, and vinegar. It increases Kapha and Pitta and decreases Vata. It is a heating taste, which counters thirst, helps maintain acidity and improves appetite and digestion. In excess, it increases acidity and causes excessive thirst, heartburn and ulcers. It is a stimulant and can energize the body as well as enlighten the mind. It is associated with the emotions of appreciation, recognition, envy, jealousy, and hate. It is made of the elements earth and fire.

Salty

The salty taste is found in sea salt, rock salt, and seaweed. It increases Kapha and Pitta and decreases Vata. It is a slightly heating taste, which maintains proper electrolyte balance and metabolism, helps cleanse the body of waste and improves appetite and digestion. It is associated with the emotions of enthusiasm, courage, greed, and addiction. It is made of the elements water and fire.

Pungent

The hot taste is found in hot spices, like peppers or ginger; also in vegetables like onions, garlic and radishes. It increases Pitta and Vata and decreases Kapha. It is drying and heating in nature, which improves metabolism, appetite and digestion, clears the sinuses, and promotes clarity of perception. In excess it causes burning, choking and sexual debility. It can also increase irritability and anger. It is associated with the emotions of vitality, clarity, anger and violence. It is made of the elements air and fire.

Bitter

The bitter taste is found in many herbs, like aloe vera, turmeric, golden seal, and coffee. It decreases Pitta and Kapha and increases Vata. It is a cooling and drying taste, which tones the organs, increases appetite, relieves fevers, and is detoxifying. In excess, it damages the heart and causes sexual debility. It is associated with the emotions introspection, self-awareness, grief and disappointment. It is made of the elements air and ether.

Astringent

The astringent taste is found in many common herbs and green vegetables, such as alfalfa sprouts, green beans, okra, and chickpeas. It decreases Kapha and Pitta and increases Vata. It is a cooling, drying taste, which reduces secretions, particularly sweating. In excess, it causes dryness, constipation, heart spasms, and thirst. It is associated with the emotions of groundedness, fear, and anxiety. It is made of the elements air and earth.

Sour, salty and pungent are heating in their effect, while bitter, astringent and sweet are cooling.

Chart 4 in appendix I summarizes the effects of taste on each of the doshas.

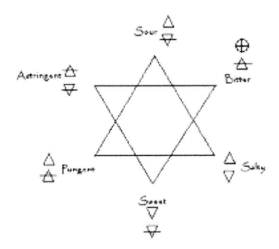

Our teeth are the mortar and pestle of the body, grinding the raw materials together with the appropriate digestive fires, provided by the saliva, into a homogeneous mass with determined qualities.

Each of our senses provides a doorway of perception to the activities of Nature and conversely provides a doorway through which our elemental balance can be affected. The senses and their associated sense organ have an elemental affinity as shown in the table below.

Element	Sense	Sensory Organ
Ether	Hearing	Ears
Air	Touch	Skin
Fire	Sight	Eyes
Water	Taste	Tongue
Earth	Smell	Nose

These natural affinities allow us to modify our elemental balance through specific pathways. By understanding our constitution, we will understand the types of therapies that will be most effective on us, providing many options for treatment through each of the senses, such as light therapy, aromatherapy, massage, etc.

What's your Dosha?

Chart 3 in appendix I will help you determine what your dosha blend is like in a very generalized way. See how it compares to the elemental balance indicated in your astrological chart.

We all possess qualities from each of the three doshas, but in different proportions.

Examine the various characteristics listed, then look across the columns to select the phrase which fits you the best. Answer the questions carefully and truthfully, not how you would like your answers to be. Also answer each question by examining that particular characteristic's influence throughout your life, not just in the present few years. When you are done, tally up the numbers for vata, pitta, and kapha. Most people show a prominent dosha and often two doshas may be equally dominant. Very rarely all three doshas are in balance; these people tend to be very healthy all the time.

There are seven combinations; vata, pitta, kapha, vata-pitta, vata-kapha, pitta-kapha, and vata-pitta-kapha. With only seven types, why aren't we all more closely alike? The different qualities that are expressed through the doshas are of differing prominence. For example, a person with vata predominant may have enhanced dryness, while another vata person may express more of the cold quality of vata. The combinations are infinite.

This little bit of self-knowledge is key to maintaining the health of your body and mind. It represents the balance of your unique blend of Alchemical Salt, Sulfur, and Mercury, and that is the balance you need to maintain in order to stay healthy on all levels of your constitution.

We naturally tend to gravitate toward foods and activities which are like our nature, and so accumulate an excess of the associated doshas. So a vata dominant person will enjoy dry crackers and chips which are also full of vata dry qualities, while a pitta predominant person will enjoy spicy hot foods. If we accumulate too much of an excess beyond our usual constitution, we become unbalanced, our digestion is impaired and toxins (ama) are formed, which can lead to disease conditions.

After you have determined your basic constitution or blend of the doshas, you can use the same form to determine how the doshas are acting in you presently. Just go through the form and answer according to conditions as they exist now. For example you may have a tendency to gain or lose weight in recent months, or a tendency to constipation, or have dry skin or loss of appetite which seems abnormal compared to your usual self.

These can serve as indicators of unbalanced dosha activity from our natural constitution and show where we can apply ourselves to reestablishing our balance.

In astrology, each of the planets, as a living being, also has its own unique blend of the doshas. How this blend of elemental qualities expresses itself is modified by the sign of the zodiac in which the planet is residing. The cool moisture of the Moon may turn into steam in Aries or ice under the influence of Capricorn.

Examine the various planets in your natal horoscope, and determine which of the essential qualities are represented. This includes qualities of the sign and its associated element as affecting the more prominent energies of the planet. These influence tone or modify the planetary energy and determine which characteristics of the planet will be increased and which will be passivated, like adding spices to our food when we cook.

If your birthchart shows a lot of water and earth qualities, that would be kapha. Fire predominant indicates pitta tendencies, while a predominance of activity in air signs will indicate vata qualities.

How our natal chart is affected by current astrological conditions can provide a valuable key allowing us to make adjustments to our daily routine which will keep us healthy, just like preparing for seasonal weather changes. Maintaining our balance amidst constantly changing conditions is the goal.

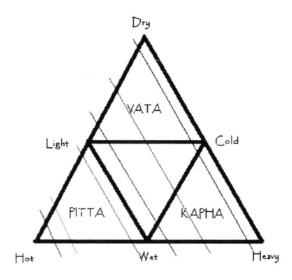

The diagram above shows common qualities the doshas share, overlaid with the spectrum of color bands and associated planets. Vata and Kapha are pacified with the warmer colors of red, orange and yellow, while Pitta is pacified with the cooler colors of blue, violet and indigo. Green is somewhat tridoshic and can pacify all of the doshas.

The opposite qualities of hot, wet, and heavy will decrease Vata, while dry, cold, and heavy will pacify Pitta, and the light, hot, and dry qualities pacify Kapha.

Mars has a lot of fiery Pitta qualities. The Sun also has a lot of pitta but shifted toward a lighter, moist, digestive fire. Saturn manifests the cold, dry, and heavy qualities of Vata and Kapha.

Materials can be conveniently classified, based on their dominant qualities, into one of the seven possible combinations of vata, pitta and kapha and thus how they will affect our doshic balance. These are sometimes referred to as their "VPK Factors".

Chart 6 of appendix I lists a number of herbs and spices and their effect on the doshas (their VPK Factors), as well as their overall heating or cooling quality. As an example, V+ indicates the material increases vata, V- means vata is decreased and Vo means the substance has no effect on vata. Herbs marked as VPK= are those which strengthen all three doshas equally.

CHAPTER FOUR

Alchemical Lifestyle

We are all different in our basic makeup; like fingerprints we share many of the same harmonics, but the strengths and weaknesses of lungs and liver for example in one will be quite different in another. Therefore, to suggest a "magic" diet for all to follow would be useless, even dangerous. I have often heard people complain of constant illness popping up when they "have been so careful" to eat only fresh, raw organic fruits and vegetables. Such a regimen was entirely inappropriate for their particular blend of energies or balance of doshas, whereas for another, the same diet helped them regain a state of excellent health. Not everyone's digestive fire works at the same intensity. If we overburden our fire, our systems get choked by the smoke of unburned food and then we get problems; you could even choke the fire out.

We usually favor food and activities which are like our dominant dosha, so a vata dominant person for example will always be on the go, love dry crackers and chips; all the things that increase vata. A kapha dominant person will love milk, cheese and ice cream, as well as naps in the daytime. All of these will increase kapha. The pitta person will pour on the hot spices, adding fire to fire. Soon we've accumulated an excess of the dosha and start to feel the effects as disease symptoms.

"Like increases like" and "opposites balance" is all we have to remember in maintaining our proper balance.

Maintaining balanced elimination is also important. You will recall from above that the formation and maturing of each bodily tissue have an associated waste product. These are called the "Mala" of each tissue. In the West, it seems to be one of those topics of little discussion, but we should keep a vigilant eye on the quality and frequency of our waste streams. The color, texture, odor and quantity of the various wastes give valuable clues to our health and balance of the elements within.

In modern society, we often adopt a tendency of repressing or postponing the urge to void wastes. We don't answer Nature's call right away, or we repress a cough, a sneeze, a burp, or passing wind, because it is inconvenient or embarrassing at the time. This can lead to problems later as blockages develop in the free flow of energy through our Nadis or subtle energy structure.

Thoughts and Doshas

We've been examining the interaction of Mercury, Sulfur, and Salt principles, or Vata, Pitta, and Kapha, largely from the physical standpoint of bodily health, the "food body". The mind and body are a whole, so the condition of one reflects the condition of the other on a variety of levels.

Mental and emotional activity associated with each of the doshas is indicated in the table below. As an example, vata predominant people tend to be hyperactive, always on the go and filled with anxieties and insecurities. They need to channel that energy into creative pursuits, be flexible and attentive to what Nature is trying to teach. Pitta predominant people need to channel their tempers into understanding their anger's true cause, being less competitive and controlling.

Dosha	Negative States	Positive States
Vata	Anxiety	Clarity
	Insecurity	Creative
	Fear	Flexible
	Grief	Perceptive
	Depression	Alert
Pitta	Anger	Understanding
	Envy	Determined
	Hatred	Confident
	Judgemental	Concentration
	Controlling	Courage
Kapha	Greedy	Love
	Attachment	Compassion
	Possessive	Forgiving
	Boredom	Cheerful
	Lethargy	Grounded

The negative states of mind and emotion keep us focused at lower vibratory rates and cloud spiritual perceptions. This will also affect the body in a negative way, obstructing vital channels, disrupting elemental balance and initiating diseased conditions. The alchemist strives to cultivate the positive aspects of thought and emotion for personal spiritual development, but also because the mental states will affect the sensitive materials worked upon in the laboratory.

Daily and yearly cycles of the doshas

The doshas have their cycles of influence on a yearly and daily scale as indicated by the following charts.

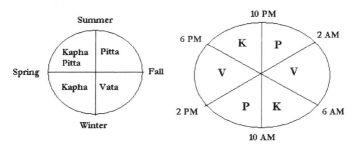

The hot summer months predominate in pitta qualities and it is easy to accumulate an excess of pitta during this time. Summer gives way to the cold dry months of autumn. The plants dry up and shed their leaves, the winds pick up and the days become colder. This is the season of vata, and exposure to these qualities can lead to an excess of vata. The cold and wet winter months predominate in kapha qualities. Everything slows and begins to break down. The moist heat of spring is a blending of the doshas pitta and kapha. The nurturing quality of this moist heat brings the rebirth of nature, and the cycle begins again as pitta slowly takes prominence and summer is upon us.

Knowing this cycling of influences, we can take measures to dress appropriately, eat the proper foods and perform the proper activities which will maintain our balance and avoid accumulating an excess of the seasonal dosha.

On the daily level, each dosha is prominent in four-hour periods. We can take advantage of this knowledge to perform

49

appropriate activities in our day-to-day lives and also within the laboratory. For example, during the hours of 10 am to 2 pm, pitta dominates and our digestive fire is at its peak. This is the time at which we should eat our largest and heaviest meal.

If we sleep in later than 6 am, we will accumulate kapha rapidly and feel heavy and slow throughout the day.

In the laboratory, vata periods are subtle and volatile, so distillations and sublimations are enhanced during these periods. Incinerations and calcinations can be performed during pitta periods, and dissolutions or crystallizations during kapha's influence.

If we can additionally work under the appropriate moon phase and planetary hour, the effects are especially enhanced.

Sattva/Rajas/Tamas—The three universal qualities of consciousness

The three Mahagunas or Great Qualities can be thought of as representing the primary modes of expression of the Absolute. In a sense, they are the divine aspect of the alchemist's Salt, Sulfur and Mercury.

In qabalistic terms they reflect the Supernal Triad of Kether, Chockmah, and Binah. At a more manifested level, they are related to the Mezla energies symbolized by the "Three Mother Letters", Aleph, Mem, and Shin. These are the three primary colors giving rise to all the colors.

These three universal qualities affect both our minds and our bodies. The doshas are a reflection of these qualities at a lower level, and just as all materials express a blending of the three doshas, so do all things express a combination of sattva, rajas and tamas. Some things are predominantly sattvic in nature, while others are more tamasic.

The essential properties of the three Mahagunas are as follows:

Sattva—The principle of equilibrium, potential energy, cognition, intelligence, understanding, and wisdom. It is consciousness and clarity of perception, giving rise to the mind and senses. The most subtle Mercury which unites and balances the powers of Sulfur and Salt.

Rajas—Principle of kinetic energy, action, responsible for all movement. The most volatile Sulfur. Rajas is said to give rise to all the activities of prana, which we will examine in more detail later.

Tamas—Principle of inertia, responsible for heaviness, slowness, sleep, unconsciousness and decay. The essence of Salt. Tamas is said to give rise to the Five Elements, earth, water, fire, air, and ether (quintessence), and their associated senses, touch, taste, sight, smell and hearing. These are the objects of perception, the most subtle energy of the five elements through which the gross elements evolve. They act as a medium bridging the "Abyss" between that which is above and that which is below, the volatile and fixed expressions of consciousness.

Although the elements are born in the womb of tamas, they each contain all three mahagunas in varying proportion. Earth is primarily tamasic, Water is tamas and sattva. Air is a balance of sattva and rajas, while Fire is intense rajas with sattva, and Ether is pure sattva.

Sattva is creative, rajas is preservative, and tamas is destructive. Tamas opposes the activity of rajas and the illuminating nature of sattva.

Sattva is the observer, rajas is the observation, and tamas is the object to be observed.

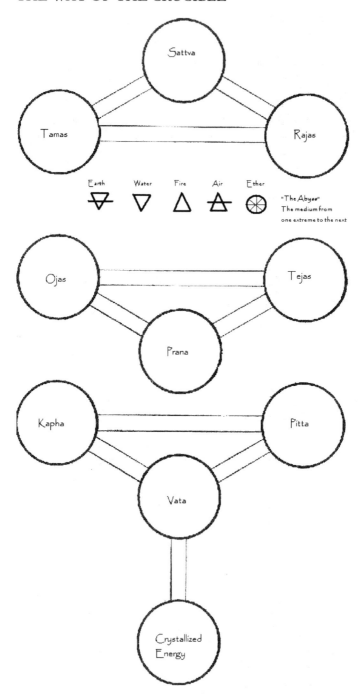

The principles are here represented in a familiar pattern. This is the pattern of the qabalistic "Tree of Life" with the philosophical principles we have been discussing, arranged by polarity and density.

Remember, these diagrams are just maps to aid our discussion; they rarely describe the whole picture, but provide seeds for insight to blossom. The map is not the territory.

Each triplicity on our "tree" acts as a functional unit. One never acts without affecting the others. This is the Triplicity inherent in the One.

> These gunas or attributes are always uniting, separating, and uniting again. Everything in this world results from their peculiar arrangement and combinations. Though cooperating to produce the world of effects, these diverse tendencies of the gunas never coalesce with each other.
> *Alchemy and Metallic Medicines,* Dash, p. 31

Prana, Tejas, and Ojas represent the sublimed essence from the doshas' action within the body (capturing, metabolizing, and absorbing food and experience). The Doshas are often thought of as the Mala (waste product) of the Three Mahagunas, acting at the physical level.

Each of the three doshas, vata, pitta and kapha, reflect a signature blend of the Three Mahagunas or fundamental characteristics of consciousness.

Vata is said to be composed of 75% Rajas, 20% Sattva, and 5% Tamas.

Pitta is 50% Rajas, 45% Sattva, and 5% Tamas.

Kapha is 5% Rajas, 20% Sattva, and 75% Tamas.

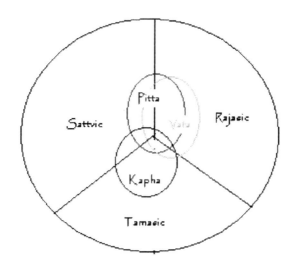

Proportions of the 3 Gunas expressed through the doshas

The three doshas are combinations of the five elements active at the relatively dense physical level. Ojas, Tejas, and Prana are their sublimed reflection active at a subtler, more energetic level. Although they are also composed of the five elements, they tend to act through the more subtle of the two elements that make up the corresponding dosha. Vata, being composed of the elements air and ether, shows its sublimed essence as Prana, working mainly through the ether element. Pitta, being composed of fire and water, has its essence as Tejas, which is active through the fire element. Kapha is water and earth; its refined essence is Ojas, operating through the water element in particular.

We mentioned Ojas earlier as the refined essence of the body, responsible for the glow of health surrounding us. Its energy is protective and nourishing with a distinctly Lunar nature. Tejas is the burning flame of pure intelligence, illumination of a Solar nature. Prana is the universal energy of life and the flow of intelligence. Another way to state this is that Tejas is intelligence, Prana is the flow of intelligence and Ojas is the medium through which the flow happens. Tejas creates Ojas and Ojas protects Prana.

At the subtle level, Ojas is the substance of the aura, Tejas is the light and color of the aura, and Prana is the activity within the aura.

This refinement of essences does not stop here. As mentioned earlier, Ojas, the pure essence of all the bodily tissues, is sublimed into Soma, the mystical Nectar of the Gods, the "Amrita" or Elixir of Immortality. Soma, like Ojas from which it derives, is related to Lunar energies, the Cosmic Plasma and the most subtle form of matter; it is the food of cells which becomes the highest expression of consciousness within us. Soma is transformed by Tejas into Prana, allowing the flow of Supreme Consciousness.

Ayurveda says that Soma is the mother of Prana and Tejas is the father of Prana. As the Emerald Tablet of Hermes Trismegistus says:

> Its Father is the Sun; its Mother is the Moon; the Wind carries it in its womb; and its nurse is the Earth.

The qualities of Ojas have been described in ancient texts as heavy, cool, soft, smooth, viscous, unctuous, sweet, stable, clear, and sticky. It has the color of ghee, the taste of honey, and the odor of fried rice. Foods such as ghee, milk, dates, almonds, basmati rice and avocados help to replenish ojas. Herbs such as ashwagandha, amalaki, and shatavari nourish ojas. Behaviors which promote the formation of ojas include yoga, meditation, reading sacred writings, chanting mantras or prayers, practicing selfless anonymous generosity and right thinking.

Remember that any quality applied to a substance will increase that quality in that substance. If a substance is exposed to qualities such as those embodied in ojas, then that substance will become more ojas-like. This includes mental impressions a substance receives.

Plants and minerals also contain their own forms of Soma, which are separated and concentrated in alchemical works in order to create powerful transformative Elixirs in the service of the alchemist.

Food and the 3 Mahagunas

In addition to its effect on the doshas, food is traditionally classified according to its effect on the body and mind, using the three Mahagunas: Sattva, Rajas and Tamas. Sattvic foods promote clarity of perception and calmness of mind. They are foods which favor spiritual growth.

Rajasic foods stimulate activity and the passions. They are generally hot, spicy foods which induce restlessness and disturb the equilibrium of the mind. Eating too fast or with a disturbed mind is also considered rajasic, as is watching TV or reading while eating. Rajasic food should be avoided by those whose aim is peace of mind, but can benefit those who generally feel sluggish and heavy.

Tamasic foods induce heaviness of the body and dullness of the mind, and ultimately benefit neither. Most meat products, canned foods, leftovers, and fermented foods are tamasic in nature. Overeating is also tamasic. The traditional advice is to fill the stomach half with food, one quarter with water, leaving the last quarter empty.

Sattva is defined as the quality of purity and goodness. Sattvic food is that which is pure, clean and wholesome. A sattvic diet is food that gives life, strength, energy, courage and self-determination. In other words, sattvic food gives us more than the gross physical requirements of the body; it also gives us the subtle nourishment necessary for vitality and clarity of consciousness. Food is a carrier of the life force or prana and is judged by the quality of its prana and by the effect it has on our consciousness.

In order to enhance spiritual development, the ancient masters recommend we follow a diet high in sattvic foods. The traditional sattvic diet is described as pure foods that are rich in prana and especially those foods which grow above ground in the solar prana filled air. Today, organic foods are recommended for both their purity and vitality. The food should be fresh as possible and freshly prepared. Leftovers of cooked food more than five hours old are considered tamasic, as the vital force begins to dissipate.

Sattvic foods are light in nature, easy to digest, mildly cooling, refreshing and not disturbing to the mind. They are best prepared with love and awareness. Remember that just as our food affects our mind, our thoughts and emotions also affect our food.

The following is a short listing of foods considered to be sattvic in nature and thus recommended as dietary mainstays for your journey along the alchemical path.

Fresh Organic Fruits: Most sun ripened fruits, including apples, apricots, bananas, berries, dates, figs, grapes, melons, lemons, mangoes, oranges, peaches, pears, pomegranates and plums, are considered especially sattvic.

Fresh Organic Dairy: We need to use the highest quality fresh dairy products to benefit from their sattvic qualities. Milk, butter, clarified butter (ghee), fresh homemade cheese, and fresh yogurt are all recommended. In fact, fresh dairy products along with fruit have been described as the most sattvic of foods, though they should never be eaten together. As dairy products become older, they tend to increase kapha and can be considered tamasic; that is why freshness is important, with the exception of ghee, which is said to become more valuable with age.

Nuts, Seeds and Oils: Fresh nuts and seeds which have been lightly roasted and salted are good additions to the sattvic diet in small portions. Good choices are almonds, coconut, pine nuts, walnuts, sesame seeds, and flax seeds. Oils should be of highest quality and preferably cold-pressed. Good choices are olive oil, sesame oil and flax oil.

Organic Vegetables: Most mild organic vegetables are considered sattvic, including beets, carrots, celery, cucumbers, green leafy vegetables, sweet potatoes and squash. Pungent vegetables like hot peppers, garlic, chives, leeks and onions are considered rajasic and vegetables such as mushrooms, tomatoes, eggplant and potatoes are considered tamasic. An excellent practice is to drink freshly made vegetable juices for their prana, live enzymes and easy absorption.

Whole Grains: Whole grains provide excellent nourishment when well cooked. Consider organic rice, basmati rice, whole wheat, oatmeal and barley. Yeasted breads are not recommended unless toasted. Wheat and other grains can be sprouted before cooking as well.

Legumes: Split mung beans, yellow split peas, organic tofu, bean sprouts, lentils and kidney beans are considered sattvic if well prepared. Other types of beans tend to aggravate vata and give excess gas. Legumes combined with whole grains offer a complete protein combination.

Sweeteners: Use raw honey (never heat or cook with honey), maple syrup, and raw sugar (not refined).

Spices: Sattvic spices are the mild spices including basil, cardamom, cinnamon, coriander, cumin, fennel, fenugreek, fresh ginger, licorice and turmeric. Rajasic spices like black pepper, red pepper and garlic are sometimes used in small amounts to keep the various channels or nadis open (rajas is used to counter tamasic qualities).

Sattvic Herbs: Other herbs can be used to directly support sattva in the mind and in meditation. These include ashwagandha, Fo Ti, calamus, gotu kola, gingko, ginseng, jatamamsi, myrrh, punarnava, shatavari, saffron, sandalwood, tulsi and rose.

Ama reducing teas

As a practical sidebar, here are a couple of simple teas for general use that are held to be useful in breaking up accumulated ama and flushing it from the system.

Take equal parts of coriander, cumin and fennel seeds reduced to powder. Simmer one teaspoonful in a cup of hot water for five or ten minutes, then strain and drink. This is also an excellent drink to settle an upset stomach.

The second recipe is to steep one teaspoonful each of powdered ginger and turmeric along with one-eighth teaspoon of black pepper in eight to ten ounces of hot water for ten minutes. Strain, then add one teaspoonful of raw cane sugar or pure maple syrup. Finally add the juice from half a lemon and enjoy.

These recipes can also be compounded as spagyric extracts with powerful effects.

CHAPTER FIVE

Importance of Water

The human body is about ninety percent water at birth and slowly drops to 50 to 65 percent by middle age, so the importance of fresh water and keeping hydrated is obvious. Quality is essential both for the body and laboratory works.

There are many wonders hidden in simple water; even its source affects its action upon the elemental balance.

Rainwater is held to be the most sattvic of natural waters and filled with solar prana. Spring water, in fact, any moving source of terrestrial water, is filled with rajasic qualities, the principle of movement. Well water is the most tamasic due to its static contact with the earth.

The amount of water we should drink is dependent on our constitution.

Vata needs most (10 to 12 cups per day), and the water should be very warm, in order to balance the cold dryness of vata. Kapha needs the least (6 to 8 cups per day); they tend to retain water efficiently, because kapha is water mixed with earth. Warm water will counter the coldness of kapha. Pitta predominant people need a medium intake of water (8 to 10 cups per day) and it should be room temperature. Cold drinks seriously hamper the digestive fire and are not recommended for anyone. They are like trying to cook when the oven door is open; the meal is poorly cooked (undigested) and leads to formation of ama in the body.

In today's world, we have to carefully look at the sources of drinking water available to us. Pollution is a global problem and pure water is hard to find. Even the pure rain, most sattvic of all, is contaminated and in some areas can become quite corrosive.

The importance and use of alchemical water works become more apparent when we consider the selectivity they offer in water qualities harmonious to the particular work.

61

One of the clearest descriptions of the alchemical work on water is found in the book *The Golden Chain of Homer*, by Anton Kirschweger, first published around 1723. You can find it on the Internet in various places and it is highly recommended for the practicing alchemist; it will provide many insights to the works of nature and open many doors to alchemical texts. Some of the practical work is also outlined in *Real Alchemy*.

One would think that distilled water is distilled water no matter how you divide the distillation products. However, experience has shown that there are subtle differences in the various fractions. Very sensitive measurements of pH can serve to indicate this, as each fraction has its own peculiarities. Many years ago, during some of my earliest experiments with water, this became dramatically clear.

At the time I was living in the mountainous region of central Idaho and had just completed the initial separation of water into the first four fractions of fire, air, water, and earth.

Each fraction had been placed into an identical glass container and sealed. It was necessary for me to leave town for several days during which the temperature had dropped to well below freezing. Upon my return, I observed the following in my water fractions: the fire of water was still liquid and had taken on a golden color. The air of water had frozen and completely exploded the glass vessel. The ice was entirely white due to a myriad of tiny air bubbles. Sitting next to this was the water of water fraction still completely liquid, and next to this was the earth of water, which had frozen into a perfectly clear piece of ice. One would have expected them to be uniformly frozen at those temperatures, but that was not the case. One water divided into different characters.

We will talk more about a particular use of the water fractions later on. At a simpler, daily use level, water provides us with a valuable tool for personal purification at both physical and subtle levels of our constitution.

Liquids in general, and water in particular, come under the rulership of the Moon. Water has a strong connection with the Moon and shares its magnetic and nurturing qualities as well as its susceptibility to psychic impression. We can utilize this knowledge of water's hidden qualities to assist our spiritual development, and our work in the laboratory.

Whereas Fire is the great purifier of the volatile energies, water is the purifier of the fixed energies.

The magnetic power of water, that is to say, its power to attract, is temperature sensitive. It is most active near the freezing point, then diminishes as the temperature rises. When the water reaches body temperature, its attractive power is almost negligible. I should mention here that water contains a portion of all five of the Elements and so can be "charged" at any temperature. The magnetic quality we are discussing here is something different. We will talk about "charging" materials later on.

If we impress thought and feeling onto water, those energies will be attracted to the water and held, like a magnet holds onto iron. In practice, each time we wash in cold running water, we can impress unwanted thoughts and emotions into the water and they will be carried away from us. This provides us with a simple way to clean a part of our subtle anatomy each time we wash our hands or take a shower, or even swim in a river.

The next time you wash your hands, collect your thoughts that you are not only washing away physical impurities, but also anxiety, depression, tiredness, and sorrow. All of these are drawn out by the attractive power of water and washed down the drain away from your physical and subtle bodies. With a little practice you will be surprised at how effective this is and pleased with newfound reserves of energy throughout the day.

CHAPTER SIX

Manipulating Chi/Prana/Mercury

We have mentioned Prana at various points in this text as the vital life force or Mercury of the alchemist and intimated its presence streaming out from the Sun and filling the air we breathe; but this is just the tip of the iceberg.

In Sanskrit, the word *prana* means "Primary Energy" and carries many levels of meaning, from the subtleties of breath to the energy of consciousness itself.

Prana is considered as the master form of all energy operating at all levels of body, soul, and spirit; even the spiritual force of kundalini, the inner power that transforms consciousness, develops from the awakened prana. The entire universe in both physical and subtle aspects is a manifestation and modification of prana, which is the original creative power.

Prana is broadly classified into two main types. There is cosmic prana, which is inexhaustible and diffused throughout the cosmos, and there is individual prana, which pervades a particular living organism. Western alchemists mention these as Universal Mercury and Determined Mercury. The first is the unmanifested aspect of prana, which is the energy of pure consciousness that transcends all creation, and the second or manifested prana is the force of creation itself.

The Mahaguna rajas, the active force of Nature, is said to give rise to the activities of prana.

Sattva by itself neither offers resistance nor does it work, and tamas always offers resistance to motion as well as to conscious reflection. When the rajasic qualities of prana unite with sattva, spiritual growth results. When the inertia of tamas is overcome by the activity of prana, ignorance, death and decay result.

The main pranic activity in the body is breathing, followed by the absorption of prana from the food we eat. The alchemists constantly remind us that control of the fire is the key to success in the Art, so at the personal level, the control of this active agent called prana is of great importance to our spiritual progress.

The control of prana is called Pranayama, a term which is often mistakenly translated to mean breath control. Pranayama is not concerned solely with breath control; breathing is only one of the many exercises through which we get to the real pranayama.

Breath, movement, and attention are all important aspects to be developed as we learn to control and direct the actions of prana both within and without. The ultimate perfection of pranayama is preformed by the mind alone.

The illustration above from an ancient Egyptian stela depicts the five aspects of prana as flowers, streaming out from the Sun.

Breathing correctly is something many take for granted or give little thought to. Mastery of the breath is a powerful tool for self transformation, so we should give it some consideration as part of our discipline along the alchemical path.

The Latin word *spiro*, meaning to breathe, is the root of the word *spirit*, and thus we see the connection with the alchemist's Mercury or life force carried by the air.

To begin at the beginning, we need to realize that the nose is the proper organ for breathing. There are many people who are in the bad habit of breathing through the mouth. The nose is specially designed to filter and warm the air before it passes into the lungs. Mouth breathing bypasses this important feature and leaves us vulnerable to a number of respiratory problems. So, if you don't already, learn to breathe through the nose, not the mouth.

Another point to keep in mind is that the skin acts as a third lung. Although it acts passively, the skin plays an important role in eliminating wastes from the body and absorbing prana from the air. To keep the skin fresh and the pores open, it is a good practice to perform a thorough dry brushing of all the skin using a natural fiber brush of medium stiffness. This helps exfoliate dead skin and adhering waste as well as open the clogged pores.

Basic pranayama exercises

You may have noticed that your nostrils do not always work at the same efficiency; often one is more clear than the other, but it changes throughout the day.

Within the body, the channels of prana include not only the respiratory tract but also the entire gastrointestinal tract, the left chamber of the heart, the hypothalamus, and of course the nose.

At a more subtle level, prana flows through all of the body's structure and particularly along three main channels called Ida, Pingala, and Sushumna. Ida is also known as the Moon channel; it moves through the left nostril and energizes the right half of the brain. The left lung has two lobes and the flow of breath here has a cooling, magnetic effect. Pingala is the Sun channel, connected to the right nostril and left half of the brain. The right lung has three lobes so when the breath moves here, it has a heating, electric effect. Sushumna is Mercury the mediator and flows through the central channel along the spine. The Central Fire of Kundalini, the transforming fire, rises through this middle channel subliming our consciousness along with it. This is the Salt, Sulfur, and Mercury of breath.

These three primary channels of the life force are often depicted as two serpents coiled upon a staff and form the caduceus, modern symbol of the medical profession. This is also called the staff of Hermes (Thoth/Mercury), recalling the myth of Thoth mediating a fight between two serpents by placing his staff between them, upon which they climbed.

The dual currents of Ida and Pingala (Moon and Sun) coil up Sushumna (Mercury/Prana), the central channel of the spine. The points where the serpents cross give rise to the nodes of consciousness called Chakras.

When the breath flows predominantly through one or the other of the nostrils, there are various activities which can be benefited by the increased energy present and its character.

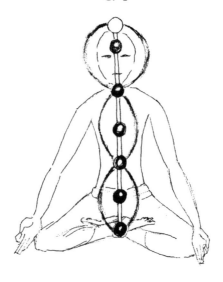

Cycles of breath

Beginning on the new moon (waxing), the breath comes into the left nostril (ida) at sunrise and stays there for two hours; after which time it switches to the right nostril (pingala) for two hours; after which it again switches to the left. This pattern continues all day and is the same for the first three days of the waxing moon. On the fourth day, the breath comes into the right nostril at sunrise and the pattern of switching nostril to nostril is the same for the next three days; after which on the seventh day at sunrise, the breath enters the left again.

This pattern continues until the full moon, when the breath enters the right nostril at sunrise. Between this constant switching from nostril to nostril, there arise periods where the breath is even between the two nostrils and said to be flowing in the sushumna nadi.

Each of the two-hour periods is further subdivided into five periods of twenty-four minutes, each of which is allotted to the flow of the five elements in this order: ether, air, fire, water, earth. Even these sub-divisions are divided, but for most practical works this is of lesser importance.

Actions undertaken during breath cycles

As the right and left brain hemispheres become activated by breath, there are definite changes in our mental states and body chemistry. When the breath is predominantly through the right nostril, the body becomes more acidic and heated. The right lung and nerve currents on the right side of the body are more active, but the left half of the brain is stimulated. This is good for all short-term activities such as hard physical labor, bathing, eating (digestion is enhanced), sleeping, reading and writing.

When the left nostril predominates, the body chemistry is more alkaline and cooled, and the left lung and left side nerve currents are activated along with the right side of the brain. This is a good time for all long term peaceful activities, including beginning new projects, practice of the arts, contemplation to gain insights on a problem, and drinking liquids.

69

When the breath is through both nostrils evenly, the current of energy flows through sushumna nadi, and only meditation and exaltation of consciousness should be attended to.

These guidelines on breath provide us with a simple means to consciously alter our body chemistry and mental states. In order to change which nostril predominates, it is usually sufficient to lie down on the opposite side and breathe only through the nostril you want to open. For example, if you wanted to breathe out of the right nostril, you would lie down on your left side, apply external pressure with thumb or finger to the left nostril to close it, and breathe from the right nostril only. Generally it takes about three to ten minutes for the change to complete.

The Total Breath

Many people breathe with only a portion of the lungs; this is called shallow breathing. Usually only the upper part of the lungs are filled and this leads to poor oxygenation of the blood, making us feel tired all the time. Proper breathing begins with the abdomen and is called the total, full, or yogic breath.

Fully exhale, through the nose, and draw the abdomen in towards the spine. On the inhale, the abdomen is pushed outward, causing the diaphragm to move downward and draw air into the bottom of the lungs. At the top of the inhalation, the chest is lifted and expanded to fill the upper portion of the lungs. A short pause at the top of the breath allows a more efficient exchange of gases and prana to occur. The out breath should be an effortless release of the air, concluded by drawing the abdomen in again in order to push out the remaining air at the bottom of the lungs. This all sounds difficult, but it is quite easy and natural in practice.

With a little conscious effort and practice, this will become your normal way of breathing and you will be surprised at the increase of energy and endurance you have during the day. If you can take the breath down below eight times per minute, the pituitary starts secreting fully. If the breath is less than four times per minute, the pineal gland starts to function fully and deep meditation is automatic.

Alternate nostril breathing

There are a number of simple breathing techniques we can use to alter our body chemistry, change our moods, clear out toxins, or accumulate prana for our alchemical works.

The technique of alternate nostril breathing balances the currents of ida and pingala along the sushumna, enhancing meditation and exaltation of consciousness.

Using the thumb of the right hand, close the right nostril and inhale through the left nostril. Now uncover the right nostril and using the ring finger, close the left nostril and exhale through the right nostril. With left nostril still closed, inhale through the right nostril, then block it again with the thumb and exhale through the left nostril. Repeat this process for three to ten minutes and sushumna should open with the breath balanced in both nostrils.

Breath of fire

A powerful method for burning toxins from the body is called the breath of fire. In this practice, the focus is at the navel point. The breath is rapid (two or three per second), continuous and powerful, with no pause between inhale and exhale.

As you exhale, the air is pushed out by drawing the navel point in towards the spine sharply. The air is drawn in by pushing the navel point out just as rapidly. The chest area remains relaxed. This is a very balanced breath with no emphasis on the inhale or exhale; both are performed with equal power. This is a cleansing breath for the blood and helps to release old toxins from the lungs, mucous lining, and blood vessels. Start slowly with one minute of practice, then work your way up to ten or even twenty minutes of practice.

Sitali or cooling breath

This breath technique is used to cool the body and is useful to regulate fevers and blood pressure; it also cures many digestive ailments. This is one of the rare times breathing

71

through the mouth is recommended. Just curl the tongue and extend the tip out past the lips. Inhale deeply, drawing the air in through the curled tongue. Exhale through the nose. If you can't curl your tongue, inhale through your clenched teeth.

Porous breathing

This final breathing method we will examine is perhaps the most useful for many alchemical experiments. An important principle to remember here is that where we focus our attention is where our energy goes. We mentioned earlier that the skin acts as a third lung. In this method we use the total breath, but instead of paying attention to the nose, lungs and abdomen, we focus our attention on the total area of our skin.

As you breathe in, feel the air and energy entering in through every pore of the skin's surface and concentrating itself evenly throughout the body. Exhale normally but feel the energy stay within you. With some practice you will be able to fill the entire body with fresh energy in just a few breaths.

You can also practice breathing in and out from specific areas of the body like a foot or hand. In addition to this, you can mentally impress the incoming energy with specific qualities. For example, let's say you have been on your feet all day and they ache. Practice breathing in and out of the feet and on the inbreath you are drawing in soothing, healing energy, while on the out breath you are releasing pain and tension out of the feet.

With practice you will be able to fill the body with energy of a specific vibration, concentrate it into your hands, and then release it into an object or substance you are working upon in the laboratory. This is a subject we will come back to later.

Body Posture

We mentioned earlier that the types of activity we perform, even the posture of the body, have an effect on our balance of energies. The physical body is the vehicle for our internal energies, which are derived from Prana. The various postures of the body, called "Asanas", are vehicles through which Prana is directed.

At the most basic level, an asana is a physical pose, a kind of bodily gesture. In practice we place the body into a position that sends a specific message and produces a specific result depending upon the shape that it creates with the body.

The way we energize the asana through Prana and our state of mind during its performance are equally important. How prana will be affected depends upon various factors including how slowly or quickly we move into the posture, the amount of force required to hold the position, and how we breathe during the whole process.

The goal of asana practice is to calm the body so that we can work on our Prana. Prana manifests most powerfully when the body is still. Each asana can enhance a particular effect upon the balance and blending of the three doshas.

A gentle, slow asana practice evenly balanced on both sides of the body is the ideal exercise for Vata types. Vatas are benefited a great deal by asana practice because it helps alleviate excess accumulated Vata. The gentle massaging action on the muscles and joints releases nervous tension and balances out the system. Keep the body calm, centered and relaxed; do the asana slowly, gently and without undue force or abrupt movements. Keep the breath deep, calm and strong, emphasizing inhalation.

Pitta types benefit from asana practices which cool down the head and blood, calm the heart and relieve tension. The body should be kept cool and relaxed; do the asanas in a surrendering manner to remove heat and tension. Keep the breath cool and relaxed. In some cases it is recommended to exhale through the mouth to relieve excessive heat as needed.

People with Kapha dominant need movement and stimulation to balance their tendency to complacency and inertia. They generally have great endurance once they get going. Keep the body light and moving, warm and dry. Keep the Prana upward moving and circulating; take deep, rapid breaths if necessary to maintain energy.

Ayurveda does not look upon asanas as fixed forms that by themselves either increase or decrease the doshas. It views them as conduits for energy that can be directed toward balancing the doshas, when used correctly. Just as individual foods have their specific effects to increase or decrease the doshas, how we prepare the food, alter its qualities with spices, or cook it is as significant as the particular foods themselves. Asanas are conditions of energy, which in turn are manifestations of consciousness. The energy and attention that we put into the pose are as important as the pose itself.

There are far too many asanas for us to do justice to the subject here, so it will be helpful to find an elementary text on Hatha Yoga as a guide.

CHAPTER SEVEN

Koshas/Chakras/The Energy Bodies

It is common to speak of our higher levels of being or deeper aspects of ourselves, so it will be useful to gain a clear picture of just what this all means. Spiritual traditions of all ages have divided our constitution into a number of distinct aspects. In alchemy, we speak of the body, soul and spirit, or salt, sulfur and mercury inherent in all things.

In ancient Egypt, a person was seen to be composed of a number of physical and spiritual components, most importantly the Khat, the Ka and the Ba. The qabalist will speak of the Guph, Nephesh, Ruach and Yechidah. Today, we hear of the physical body, etheric body, astral body, mental body, and spiritual body.

These are all states of the one energy or consciousness; mind seems to manifest itself in different layers. As the Emerald Tablet of Hermes says, "all things are derived from the one only thing". It is sometimes difficult to distinguish where one layer ends and another begins because all of Nature is really a continuum of energy separated only by relative rates of vibration. However, we find it convenient to divide Nature into a number of levels or spheres of being in order to recognize the various modes we perceive. Emotions have their levels of vibration and thoughts have their levels. In a sense, the physical body is crystallized mind.

Plants and minerals have their own energy bodies as well.

Most of these systems of division make comparisons or refer their roots back to the system derived by Indian sages, which provides a concise map of our constitution.

In essence, the core of our being, pure consciousness, expresses itself through three bodies wrapped in a series of sheathes called Koshas. This is often described as like the layers of an onion enclosing the divine spark at its center, our innate or central fire.

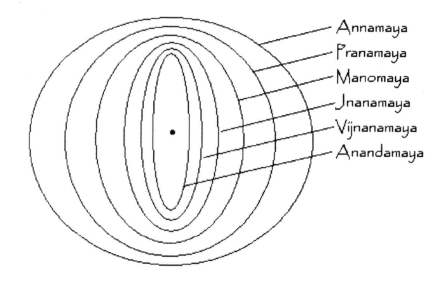

Annamaya
Pranamaya
Manomaya
Jnanamaya
Vijnanamaya
Anandamaya

Koshas or Sheaths of the Body

Starting with the most dense and lowest vibratory level is the Annamaya Kosha. Anna means food and maya means "made of", so this is often called the "Food Body"; it is the physical vehicle nourished by food and drink. Our diet and the physical components of alchemical medicines affect this level.

The term Maya is often translated as illusion, but it is not so much illusion as we normally think of that word but relative reality as opposed to absolute reality. Maya describes the principle of ever-changing matter, constantly going through the cycles of creation, preservation and destruction, so it is never static. What we perceive is constantly changing and thus a transitory illusion.

The second level is termed the Pranamaya Kosha, the vital body which sustains the activities of prana. This sheath is the first layer of the subtle body; it connects the higher and lower aspects of our constitution. In the West, this is referred to as the "Etheric Body". The solar prana in the air we breathe and the food we eat, and that captured in alchemical preparations vitalize this part of our constitution.

The third kosha is called Manomaya Kosha. This is from the term Manas, which means mind. This portion of the subtle body is responsible for psychic activities and emotional states. In the West it is called the astral body.

Remember that with each ascending level, the rate of vibration is increasing. The activity of this level is closely connected to the activity of Pranamaya Kosha, also part of the astral body. The subtle psychic impressions captured by or impressed upon alchemical medicines during their preparation affect this kosha and the next two.

The fourth kosha is called Jnanamaya Kosha, from the word meaning knowledge. This sheath along with the next, form the mental body. At this level, what we might term the lower mind has its seat.

The fifth level is known as Vijnanamaya Kosha and is the seat of higher intellectual activities.

At the sixth level, surrounding the very core of our being, we find Anandamaya Kosha, the Body of Bliss responsible for all spiritual activities. Possessing a very high rate of vibration, it is the Spiritual Body or Causal Body. Hidden within this sheath is our spark of divine consciousness, our Quintessence and Central Fire. The Soma of plants and minerals concentrated within alchemical preparations affect this level to fortify the incorruptible body of light.

While it is convenient to think of these several levels as distinct entities, remember they shade into one another, they all interconnect and affect each other.

Each kosha is anchored to a chakra or energy node located along the spine as the following table describes:

Kosha	Chakra	Location	Function
Annamaya	Muladhara	Base of Spine	Instinct
Pranamaya	Svadhisthana	Abdomen	Vitality
Manomaya	Manipura	Solar Plexus	Will
Jnanamaya	Anahata	Heart	Emotional Love
Vijnanamaya	Vishudda	Throat	Communication
Anandamaya	Ajna	Third Eye	Divine Sight
Beyond Koshas	Sahasrara	Crown of Head	Illumination

The chakras, along with their associated nadis, Ida, Pingala, and Sushumna, are the major constituents known as the Channels of Mind or Mano Vaha Srotas. From muladhara chakra to sahasrara chakra; from annamaya kosha to anandamaya kosha is a journey of mind. For every change in consciousness, there is a corresponding change in vibration of matter and vice-versa.

When the mind is working through the root chakra at the base of the spine, it works for food, survival and position. When the mind works through the second chakra (svadisthana) it is concerned with self-identity and procreation. A mind operating through the third chakra is concerned about power, ambition, competition and control. These three lower chakras belong to the animal nature and operate at a relatively low vibratory rate. Alchemy is about "raising the vibrations" and learning to operate from the higher centers of being.

As we clean out the body, it has a beneficial effect on the higher koshas; as we learn to control breath, it also has beneficial effects on the other koshas; and as we learn to control the thoughts and desire nature, the effects run through all of the koshas.

Then we begin to operate through the higher centers of consciousness until we realize our divine nature which resides at our core.

The alchemical process involves working on all of these levels, bringing them to their highest degree of perfection and highest rate of vibration.

> For you now have a Heavenly Oil that shines on a
> dark night and emits light as from a glowing coal.
> And the reason for this is that its innermost

power and soul has become thrown out unto the outermost, and the hidden soul is now revealed and shines through the pure body as a light through a lantern: Just as on Judgement Day our present invisible and internal souls will manifest through our clarified bodies, that in this life are impure and dark, but the soul will then be revealed and seen unto the outermost of the body, and will shine as the bright sun.

Roger Bacon, *Tract on the Tincture and Oil of Antimony*

In the next chapter, we examine a methodology for effecting such powerful changes in our constitution by introducing the concept of "Rasayana", which is of particular relevance to the laboratory works that follow later on.

The rasayana or rejuvenation therapy is meant to provide the required nourishment to all these six layers of the individual and not merely the physical body.

Dash, *Metallic Med.*, p. 22

CHAPTER EIGHT

Science of Rasayana

Rasayana is a word for rejuvenation in Ayurveda. Rasayana is a branch of Ayurveda and a development from Rasa Shastra which most closely mirrors the alchemical tradition of the West. It literally means "the path of juice", which aims to nourish, restore and balance the body functions. Rasayana is used to rejuvenate the general health of the body and achieve the individual's maximum potential. This potential is based on the constitutional type of each person. Therefore, if an imbalance or a disease process occurs within the body, it is possible to recover from this imbalance and regain equilibrium, halt deterioration or premature aging and reinstate order in all of the bodies. This restoration of balance or potential within the human economy by maintaining or replenishing the "juice" is called Rasayana Therapy.

According to classical Ayurvedic texts, Rasayana therapy arrests aging and enhances intelligence, memory, strength, youth, luster, clarity of voice and vigor. It works to nourish each of the seven tissues (dhatus), thus preventing degenerative changes and illness.

Rasayana is held to improve metabolic processes and produce the highest quality bodily tissues in order to eradicate senility and other diseases of old age. From this treatment, one attains optimal strength of physique and clarity of the senses. It also builds natural resistance against infection.

One of the ways by which rejuvenation may be attained is through the use of specially prepared alchemical medicines. Rasayana medicines form a group of medicinal herbs and metallic preparations that are often employed as complex mixtures which bring about a striking improvement in both the mental and physical health of the individual. These preparations not only work on the physical body but also open the subtle channels (nadis) of man's finer structure and increase the production of refined ojas and prana.

81

The following list of herbs (mostly from India) and their actions includes some of the more common ones used for rasayana therapy:

Bala *(Sida cordifolia)*, Kashmari *(Gmelina arborea)*, and Ashwagandha *(Withania somnifera)* act as general dietary aids for increasing the strength of tissues and organs.

Pippali *(Piper longum)* acts as "Respiratory Rasayana". Its use in several preparations is also mentioned by Paracelsus.

Garlic *(Allium sativum)* increases various tissue enzymes and enhances digestion. It has long been prized for its medicinal powers, but the overuse of garlic can increase pitta and it is tamasic in nature, so use wisely.

Haritaki *(Terminalia chebula)* acts as "Digestive Rasayana" by eliminating waste products from the tissues and organs, particularly in the intestinal tract.

Brahmi *(Bacopa monnieri)*, Gotu Kola *(Hydrocotyl asiatica)*, Vacha *(Acorus calamus)*, Shankhpushpi *(Convolvulus pluricaulis)* and Fo Ti *(Centella asiatica)* all help to increase intellect and memory, thereby acting as Brain Tonics.

Guggulu *(Commiphora mukul)*, a close relative of Myrrh, helps further digestion of waste products and removal of Ama. This resinous material is used in many combinations to enhance their effects.

Amalaki *(Embelica officinalis)* and Guduchi *(Tinospora cordifolia)* decrease the catabolic process and thus postpone ageing. They are known as Adaptogenic tonics.

Bhallataka *(Semicarpus anacardium)* has a powerful immune promoting substance held to overcome many conditions like rheumatoid arthritis and some stages and types of malignancies (cancer).

Bhringaraj *(Eclipta alba)* prevents aging processes and helps rejuvenate bones, teeth, hair, sight, hearing and memory. It also acts as a powerful liver rejuvenator.

Punarnava *(Boerrhevia diffusa)* promotes the functions of the kidney and improves the regenerating capacity of the nephrons (a functional unit of the kidneys).

Shatavari *(Asparagus racemosus)* is perhaps best known as a female rejuvenative. It is useful for infertility, decreased libido, threatened miscarriage, menopause, and leucorrhea.

Kapikachhu *(Mucuna pruriens)* and Ashwagandha *(Withania somnifera)* promote generative activities in the tissues, restore senile sexual dysfunctions and cure impotency.

Chyawanprash is an ancient and still very popular rejuvenating tonic of Ayurveda. It is made by mixing the elixir of eighteen roots and herbs in fresh Amalaki Paste and sugarcane syrup. Chyawanprash enhances body metabolism and immunity and is considered to be one of the most health promoting products of Ayurveda.

Other commonly used herbs include Aloe Vera, Black Pepper, Ginseng, Comfrey Root, Saffron, Elecampane, Licorice, Marshmallow, Myrrh, and Spikenard.

The combination of Saffron, Aloe, and Myrrh forms the basis of the "Elixir of Property" of Paracelsus and is also mentioned in the Circulatum Minus of Urbigerus as "a most excellent cordial, and almost of as great efficacy and virtue as the Universal Elixir itself in curing all curable distempers".

The "Primum Ens Melissa" is another herbal rasayana of the Western tradition. Its preparation was outlined in *Real Alchemy*.

The imbalances in the three doshas of vata-pitta-kapha occur sometimes by the mind and sometimes by the body's dhatu (tissues), waste products (mala), or accumulated ama (toxin deposits). Hence, rasayana therapy delves deeply into ascertaining the root cause of the illness. Then a suitable treatment is recommended to bring mind and body back into balance. Though the process might give an impression that the treatment is meant solely for the physical body, there is in fact a strong impact on the subtle levels of a person as well.

To prolong the youthfulness of the body, rasayana applies several physical and mental disciplines along with the specialized medicinal preparations, to rebuild the body's cells and tissues only after an initial process of detoxification.

Rasayana therapy increases the life force, ojas and immunity of a person and thus there is a regeneration of cells and tissues in the body.

Literally, *rasayana* means the augmentation of rasa, the vital fluid produced by the digestion of food and concentration of prana. It is the rasa flowing in the body which sustains life. Rasayana in ayurveda is the method of treatment through which the rasa is maintained in the body at its highest level of perfection.

Rasayanas prepared from medicinal plants have been used from time immemorial and have been instrumental in giving long, disease-free, vigorous lives to their users.

In addition to herbs, rasayana therapy also employs various metals, minerals, and jewels as rejuvenatives. These elements are turned into calxes or bhasmas for administration in different diseases. Calxes of gold, silver, copper and iron are widely used. We will examine the preparation of some of the bhasmas later in the practical works.

Gold and diamond, two of the most powerful and highly prized rasayanas, are specially prepared to increase ojas (the essence of the body).

Jewels or "ratnas" include precious and semiprecious stones, which are used as drugs because of their therapeutic properties. Major jewels or maharatnas include diamond, ruby, pearl, topaz, sapphire, emerald, cat's eye, and zircon. Uparatnas or minor jewels like sunstone, moonstone, lapis lazuli, jade, carnelian, and crystal are also used.

The calx of diamond is a powerful cardiac tonic and one of the best elixirs. It is applied in curing diabetes, urinary troubles, and anemia. Because of its powerful aphrodisiac qualities, the calx of diamond is used to treat impotence. The calx of rubies is used to increase intelligence, virility, and longevity. It also cures disorders produced by the vitiation of the three doshas.

Coral *(Corralium rubrum)*, calcined bones, and calcined eggshells provide calcium essential for bone growth and development as well as certain enzymatic processes. The use of coral preparations is also found in many Western alchemical texts.

The concepts of rasayana closely mirror the Western alchemical tradition concerning the "Elixir of Life". In practice, the individual must strictly follow the diets and other physical and mental regimens which constitute a powerful detoxification process before undertaking the rejuvenation process. Rejuvenation is done with the help of specific medicinal formulations and a specific lifestyle regimen dependent on the individual's constitution.

Ayurvedic literature also specifies rasayanas for each dhatu and this provides us with the opportunity of creating a set of "Seven Basics" with the goal of regeneration therapy in mind. Details are provided in chart 7 of appendix I.

Anupanas

The effectiveness of an alchemical medicine is dependent on a number of factors including the nature of our basic constitution and its relation to the particular medicine, and in the timing or when the medicine is to be administered.

Another factor which is often overlooked is the medium or vehicle in which the medicine is applied or ingested. In ayurveda, this is called the Anupana.

Most often an alchemical medicine is in the form of an alcohol or water and alcohol mixture which is placed directly on or under the tongue. Many of the old texts recommend using red wine as the vehicle. The choice of vehicle can have dramatic effects on how and where a material will act within the bodies.

The anupana can enhance the therapeutic effects of a substance, relieve side-effects, exhibit a catalytic effect which carries the material into deeper, more subtle tissues, and change the dosha the material works upon.

Here is a list of commonly used anupanas and their effect on the doshas:

Anupana	**Dosha which is reduced**
Ghee	Pitta
Sesame oil	Vata
Honey	Kapha
Hot Water	Vata and Kapha
Cold Water	Pitta
Pomegranate	Pitta

Water conveys the effects to the plasma, while honey conveys the effects to the blood and muscle tissues; it also provides a nourishing effect. Milk is often used to convey the effects to both plasma and blood while increasing the tonic effect as well.

Raw sugar also increases the tonic effects upon plasma and blood while reducing heat, protecting tissues, and aiding general metabolism.

Alcohol or red wine carries the effects into the marrow and nerve tissue as well as the more subtle tissues of the body.

The Water Work of alchemy as presented in *Real Alchemy*, wherein rainwater is distilled into twelve fractions each with an astrological correspondence, provides the operator with a set of powerful anupanas which can effectively carry a medicine to specific areas of the body or mind. They can also be used to alter the effect of planetary medicines by combination with a particular sign. For example, the stimulating effects of a Mars ruled medicine could be enhanced by administering it in the Aries fraction of water, or become more grounded by using the Leo fraction. The effects of a Venus ruled medicine could be directed more powerfully to the kidneys using the Libra water fraction or to the skin using the Capricorn fraction. The possible combinations are endless, and present the operator with a means of preparing a customized medicine for the individual based on their constitution and astrological affinities.

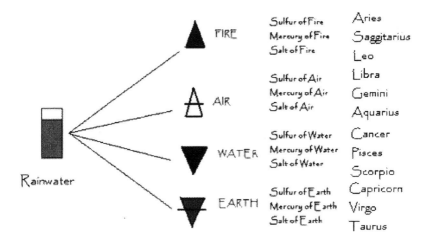

CHAPTER NINE

Qabala and Alchemical Eucharist

As we have seen, the practice of alchemy is more than just a manual operation designed to prepare a medicinal substance. It requires the participation of mind and body aspiring toward the spiritual height.

In ancient Egypt, the veil between the worlds was much thinner, so the care and maintenance of one's physical and spiritual economy were constantly in mind. The preparation of a medicine, even the preparation of a meal, was always accompanied by the use of prayers, incantations and spells.

These helped to focus the mind and intention of the operator, but also awakened subtle forces within the subject matter, stirring them into action by a type of resonance. This focusing of intention was also applied during the administration of a medicine, creating a harmony between the substance and patient, physically and energetically.

The world of the Gods constantly interacted with the world of man. These ideas became incorporated into the mythology, stories, and sciences of the time.

In the great battle between Horus and Set, Set plucked out the eye of Horus and tore it into pieces. Later, the alchemist god Thoth/Hermes used his Art to reunite the pieces and make the eye whole again. This is the Solve et Coagula in alchemy. The Eye of Horus was referred to as the "Whole Eye", and became a symbol of total health. The priest-physicians used hieroglyphic representations of pieces of the eye to denote fractions in medical prescriptions. The fractional components of the prescription would come together as the Whole Eye.

The modern symbol for a prescription, Rx, is derived from the Eye of Horus.

The "wholeness" represented by the eye referred not only to the physical body, but to the more subtle structure as well. The Eye, as the organ that perceives light, is also a symbol of spiritual perception.

The ancient mystery traditions worldwide describe the division of The One into two contending forces: the powers of light and the powers of darkness locked in eternal battle; the war in heaven between good and evil, between Cosmos and Chaos. One is the evolutionary wave, struggling to transform Chaos into order and intelligence, while the other strives towards entropy, the increase of disorder, unconsciousness, and randomness. Between these two poles exists a continuum of energies, intelligence, and substance. Within this continuum arise various levels, worlds of their own, supporting intelligence and substance appropriate to that level. The human dimension is only a small portion of that continuum, similar to the visible spectrum of light within the whole electromagnetic spectrum.

The systems of Natural Magick which became popular in the Middle Ages divided these planes of existence into four major categories, each dependent on the others and each possessed of differing levels within itself. The highest, most spiritual level was known as the Intellectual World. This represented the level of primal archetypes, the blueprints of all that which follows as the Absolute condenses into matter. In the Qabalistic tradition, this is the world of Atziluth, the divine realm. The levels within this world represent the attributes of the Gods and Goddesses of the world's religions. Indian alchemists call this Sivaloka, the Causal Plane.

The next level is less rarified and known as the Celestial World. This is the Creative World or Briah of the Qabalist, and the upper planes of Devaloka of the Indian philosophers. This is where the blueprints of the archetypal world are framed in the matrix of time and space. This is the realm of exalted spiritual beings and Archangelic powers.

Below this is the Sub-lunary or Elementary World, the realm of Yetzirah of the Qabalist, also referred to as the Astral Plane or the lower planes of Devaloka. This is the level governed by Angelic powers.

The influences of these higher worlds finally condense or crystallize into the physical world. This is the world of Assiah of the Qabalist, Bhuloka of the Indian alchemists. The vibration of the creative force has finally slowed down enough to be in the range our senses can perceive.

Remember this is all a continuum of energy, separated only by relative rates of vibration.

Throughout all time and across the world, there have been special "Words of Power" or Mantras, held to resonate with these various levels and awaken their activity in the physical world.

Such words or mantras contain seed syllables reflecting cosmic creative vibrations. These vibrations in turn condition the expression of prana, affecting both substance and mind. The word *mantra* derives from *manas*, Sanskrit for mind. The use of mantra during alchemical operations provides a means of focusing intention while at the same time impressing the material's prana or Mercury with specific characteristics.

There are mantras, in many languages, for each type of cosmic formative force and mantras for each of the planets, so the subject is beyond the scope of this book to do it justice.

The practical alchemist should investigate the subject of mantra and select those compatible with their religious or philosophical ideals. A simple use of the universal mantra, "OM", can lead to surprisingly enhanced effects.

Charging an Elixir

During each step along the path of preparing our alchemical products, the effects of our thoughts and attitude can have a powerful effect upon the substances involved. The final product will retain much of this energy, but there are methods we can apply which amplify their effectiveness. These methods are collectively referred to as "charging" the material. Much like a magnet or a battery is charged by applying the appropriate type of force, our alchemical products can be impressed with specific types of mental, spiritual, and astrological forces which help to guide the substances toward desired effects.

There are many ways to charge our elixirs, from the simple act of prayer or reciting mantras to elaborate ceremonial ritual similar to the Catholic Mass or Rites of Initiation. Here, we will only examine a few of the more commonly applied methods, and you can adapt them for use in the particular path you follow.

The simplest method is to formulate in your mind what your intentions for the elixir are to be, such as healing a specific ailment, or bringing insight to some aspect of natural law. Take a few moments to compose yourself in meditation, then with hands outstretched over the elixir, speak a short prayer or spell which expresses your intention, like saying grace at a meal, and feel the force flow out of your hands into the elixir. Now consume the elixir immediately as this type of charge tends to dissipate quickly.

Another powerful method of charging the elixir depends upon a little skill that takes some practice to acquire, called Transference of Consciousness. In this method we project a portion of our awareness into the material to be charged; in a sense, we become the material and perceive the world around us from its perspective.

You can practice transferring consciousness into simple objects in order to develop the skill.

Begin by selecting an object nearby, say a vase or small statue, into which you will project your consciousness. Now compose yourself, sitting comfortably in meditation with eyes closed. Look inside yourself to find where your consciousness is centered. This is the point inside where you feel "I am here". Some feel it in their heart area or solar plexus, while others feel as though they are centered in the throat area or between the eyes. Gather yourself at this point, using the porous breathing technique to feel as though you are a tiny sphere of light, breathing in and out slowly from this point. See and feel yourself, from this point of light, drifting out of your body, across the room and into the object you have chosen.

Try to sense the contours of this object as though it was your new body, look around the room in your mind's eye and see things from the object's point of view. After a few minutes of this exercise, gather yourself into a point of light again, breathing spherically from this center. Now see and feel yourself as this point of light, drifting back into your body to the point from which you originally left. Take a few deep breaths in the normal manner, stretch your hands and arms, as you reestablish contact with your body, then slowly open your eyes.

With a little practice, you will acquire the knack of doing this transference easily.

For charging elixirs or other materials, transfer your consciousness into them and from this new body, use the method of porous breathing to accumulate energy just as you did in your own physical body. Impress your intentions on the accumulated energy, and clearly state that the energy shall remain in the material until it has accomplished its purpose.

Now gather yourself to your point of consciousness and transfer back into your own body, leaving behind the accumulated charge of energy. This method of charging allows you to combine additional sources of energy while you are "within" the elixir, such as accumulating appropriate colors or vibrating appropriate names and words of power.

This forms the basis of one of the Qabalistic methods for charging, wherein the operator centered within the elixir accumulates the Atzilutic color of the particular planet and charges the elixir in the associated divine name. This is followed by accumulating the associated Briatic color and charging in the archangelic name, then the Yetziratic color and angelic name, and finally the Assiatic color and charging in the names of the intelligence, spirit and physical planetary names.

All of these various forms of charging an elixir are a type of Talismanic Magic, where the elixir itself is the talisman. In yet another sense, throughout its preparation the elixir is passed through a series of mystical initiations, raising it from "a wanderer in the outer darkness" into an evolved, spiritual being in the service of the alchemist.

Depending on one's religious or philosophical traditions, these ceremonies can become quite elaborate.

A simplified version used by a number of modern operators employs the use of the traditional magical Planetary Seals (chart 8 in appendix I) as a focus for invoking celestial energies. These lineal figures and associated polygons are held to place a stress on the subtle energy planes which attract their related types of energy, similar to a dialing sequence on your telephone.

Working within an established sacred space which is ritually cleansed and purified, a pentacle is placed on the altar. This pentacle can be made from a disk of clay and represents the material plane focus. On the top of the pentacle is placed the planetary seal for the planet one is working with. Ideally the planetary seal is etched into a plate of the appropriate planetary metal cut into its corresponding polygon shape. As an alternative, colored paper and ink can be used to create the seal. To enhance the effectiveness of this type of seal, you can impregnate the paper with a tea or tincture of the appropriate planetary herb, dry it and then draw the seal upon it. On top of the seal is now placed the elixir to be charged. Using the methods of invocation, ritual gesture, and/or sacred dance, the planetary energy is called forth and focused into the "stack" upon the altar. When completed, the elixir is wrapped in silk to keep the charge from dissipating. Finally, the invoked energies are released from the circle and returned to the cosmos with thanksgiving.

A passive form of this charging allows the "stack" to remain exposed during a time when the specific planetary influence is at a peak intensity. After the exposure, the elixir is wrapped in silk until used. This type of charging usually requires several exposures at optimum times to be effective.

Use of colloidal gold

In order to stabilize the charge on an elixir and to hold the charge for longer periods, the use of finely divided gold is recommended. Just a small trace of gold is all that need be added to the elixir prior to charging.

Although we will talk about preparations of colloidal gold later on, a simple method given here will be found satisfactory for most uses. The preparation is easy, and requires only a small amount of gold. This can be from a broken piece of jewelry, a piece of gold chain, or an old ring or earring. It should be at least 18 karat gold. Using pliers, hold the gold piece over a flame, even a candle will work, but be sure to hold it above the flame so it won't get sooty. Heat the gold until it is quite hot and then drop it into a small amount of water, like half an ounce. Recover the gold and repeat this process of heating and quenching ten or twelve times. Allow the liquid to stand several hours, then decant the clear liquid away from any solids at the bottom, into a dropper bottle for use. Each time the gold was heated and quenched, small particles of the gold were blown off and suspended in the water. The solution looks clear because the particles are too fine for the eye to see. You can use a laser pointer to confirm that you really have a colloidal suspension of gold particles. You should see the laser beam light up as it passes through the liquid, whereas if you pass the beam through untreated water, it passes through invisibly.

To use, simply place two or three drops of the gold suspension into the elixir you wish to charge, then proceed with any charging method you have chosen.

In each of these methods of charging, we aspire high into unseen spiritual levels to contact a pure form of the spiritual light, then progressively draw it down the planes into a physical manifestation. Like Prometheus bringing fire down from the heavens, it is the gradual condensation of Celestial Fire into our material subject.

Although written over four hundred years ago, the advice of the alchemist Basil Valentine still rings true for those who follow the Hermetic Art.

In this Meditation I found that there were five principal Heads, chiefly to be considered by the wise and prudent Spectators of our Wisdom and Art. This first of which is, Invocation of God. The second, Contemplation of Nature. The third,

True Preparation. The fourth, The Way of Using. The fifth, Utility and Fruit. For he, who regards not these, shall never obtain place among true Chymists, or fill up the number of perfect Spagyrists. Therefore touching these five Heads we shall here following treat, and so far declare them, as that the general Work may be brought to light and perfected by an intent and studious Operation.

1. Invocation of God must be made with a certain Heavenly Intention, drawn from the bottom of a pure and sincere Heart, and Conscience, free from all Ambition, Hypocrisy, and all other Vices, which have any affinity with these.

2. Next in order after Prayer is Contemplation, by which I understand an accurate attention to the business it self, under which fall these considerations first to be noted. As, what are the Circumstances of any thing, what the Matter, what the Form, whence its operations proceed, whence it is infused and implanted, how generated by the Stars, conformed by the Elements, produced and perfected by the three Principles.

3. Next to the Theory, which researcheth out the inmost properties of things, follows Preparation, which is performed by Operations of the hands, that some real work may be produced. From Preparation ariseth Knowledge, viz. Such, as opens all the fundamentals of Medicine. Operation shews how all things may be brought to light, and exposed to sight visibly: but knowledge shews the practice; and that, whence the true Practitioner is, and is no other than confirmation: because the operation of the hands manifests something that is good, and draws the

latent and hidden nature outwards, and brings it to light for good.

4. After Preparation, and especially after separation of the good from the evil, we are to proceed to the Use of the weight or dose, that neither more, nor less than is fit, may be given. For above all things, the Physician ought well to know, whether his Medicament will be weak or strong, also whether it will do good, or hurt, unless he resolve to fatten the church yard, and with the loss of his fame, and hazard of his own soul.

5. After the Medicament is taken into the body, and hath diffused it self through all the Members, that it may search out those defects against which it was administered, the Utility comes to be considered; for it is possible that a Medicament diligently prepared, and exhibited in due weight, may do more hurt than good in some Diseases, and seem to be Venom rather than Medicine. Hence an accurate reflexion is to be made to those things, which profit or help; and they are diligently to be noted, that we may be mindful to observe the same in other cases.

Triumphal Chariot of Antimony

CHAPTER TEN

Of Minerals and Metals

Food, water, and fresh air are the first line in balancing our constitution and separating the pure from the impure.

The herbal and water works of alchemy continue the process of cleaning the body tissues, opening the channels and sublimating vata, pitta, and kapha into their essences of prana, tejas, and ojas.

The mineral world represents tremendous power locked up within crystalline matter. Mineral and metallic medicines have always represented the most powerfully effective agents for the perfection of mind and body.

> Many a man kicks away with his foot a stone which would be more valuable to him than his best cow, if he only knew what great mysteries were put into it by God by means of the spirit of Nature.
>
> Paracelsus/Hartman, p. 303

Crystals

Modern science often describes matter as "frozen light" or "standing waves" of energy and the quality of mass as a phenomenon of light rays resonating back and forth, freezing themselves into a pattern. When that frozen form has a unique pattern of atomic alignment such as a crystal structure, the specific energetic signature is greatly accentuated. Metals and minerals possess intelligence; they grow crystal forms that share a unique symmetry in their atomic lattice, structuring billions of atoms in geometric alignment. This allows for a greatly amplified flow of specific energies that can harmonize and strengthen corresponding radiations that the Sun, Moon and planets shower upon the earth and within our Microcosm.

Alchemy of the East and West teaches that everything is composed of "rays" that create, maintain, and destroy the Universe. What we call matter is a slower vibratory condensation of Cosmic or Supreme Consciousness, the modifications of prana streaming out from the Sun.

Metals and gems as the flowers of the mineral kingdom are especially unique with a transmittable life force and consciousness locked within their geometric and determined crystal habit.

Although these "standing waves" we call atoms are complex, they have a spherical body of influence around them, so it is easy to picture them as little spheres. There are only two ways in which these spheres can pack themselves together, and these are called "cubic close packing" and "hexagonal close packing". The illustration below depicts the two forms.

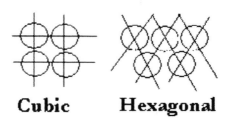

Cubic   Hexagonal

Cubic and Hexagonal close packing of spheres give rise to seven crystal systems with 32 variations. Each of the seven crystal systems has its correspondence to one of the seven ancient planets through specific modes of resonance and energy density. Thus, the crystalline form of a material can provide information concerning the type of planetary energy it conducts.

One such system of correspondence in use by modern practitioners comes from the pioneering work of the French alchemical group, The Philosophers of Nature, as indicated in the following table.

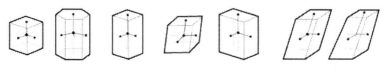

Cubic	Hexagonal	Tetragonal	Rhombic	Orthorhombic	Monoclinic	Triclinic
Saturn	Moon	Jupiter	Mercury	Mars	Sun	Venus

100

Another way of looking at the crystalline energetics is the correspondence of their form and qualities in comparison to the seven combinations of the three doshas. The cubic system, as the most perfect crystal form, reflects the balance of vata/pitta/kapha, while the hexagonal system with its rounded forms reflects the qualities of kapha. The most irregular crystal form, triclinic, reflects the erratic qualities of vata, and the sharp, thin lines of the monoclinic system show vata/pitta influence. The following table suggests these connections.

Doshas and crystals

Crystal System	Dosha Blend
Cubic	Vata/Pitta/Kapha
Hexagonal	Kapha
Tetragonal	Pitta
Rhombic	Pitta/Kapha
Orthorhombic	Vata/Kapha
Monoclinic	Vata/Pitta
Triclinic	Vata

The Chemical Sky

All materials are a reflection of celestial formative forces. Through the Law of Polarity, these forces are both visible and invisible in nature. The range of materials used in alchemical works are sometimes referred to as "The Chemical Sky". This is a translation of celestial forces into material forms. Some of the most important mineral and metallic correspondences for practical works of self transformation are shown in the table below.

SIGN	PLANET	METAL	CELL SALT
Aries	Mars +	Iron	Potassium Phosphate
Taurus	Venus -	Copper	Sodium Sulfate
Gemini	Mercury +	Mercury	Potassium Chloride
Cancer	Moon -	Silver	Calcium Fluoride
Leo	Sun +	Gold	Magnesium Phosphate
Virgo	Mercury -	Mercury	Potassium Sulfate
Libra	Venus +	Copper	Sodium Phosphate
Scorpio	Mars -	Iron	Calcium Sulfate
Sagittarius	Jupiter +	Tin	Silica
Capricorn	Saturn -	Lead	Calcium Phosphate
Aquarius	Saturn +	Lead	Sodium Chloride
Pisces	Jupiter -	Tin	Iron Phosphate

Cell salts

One way to provide yourself with a safe, easy introduction to the power of mineral and metallic medicines is through the use of Schuessler Cell Salts in homeopathic dilution.

In 1873, a German biochemist named William H. Schuessler found that there are certain essential minerals that the body requires, in proper balance, in all of its cells. An imbalance or a lack of any of these minerals may lead to disease in those tissues. Providing the missing minerals to the tissues corrects that imbalance, and so eliminates the illness. This simple system of healing has found great practical applications in maintaining health. Schuessler recognized twelve important cell salts and that by using combinations of these simple minerals, it is possible for any individual to treat themselves, simply and effectively, for a great variety of everyday minor ailments. The Schuessler cell salts may be seen as raw materials for the body, and cell-salt treatment replenishes something the body lacks and needs.

The minerals must be in a special form to be most useable to the body. Most Schuessler cell salts are in a homeopathic potency, which uses minute quantities of a substance with great effect. The homeopathic potentiation process multiplies the essential energy of a substance while at the same time decreasing the dosage required. The subtlety and gentle effectiveness of such preparations have been clinically proven since the beginnings of homeopathic treatment in the 18th century.

In the mid-nineteenth century, the work of Dr. G.W. Carey established a correspondence between the 12 cell salts and the 12 signs of the zodiac. This has provided medical astrologers with a valuable key to maintaining an individual's health.

There are several ways in which cell salts are applied therapeutically. First is through knowledge of the individual salts and their particular healing virtues. This is the common homeopathic practice and does not require knowledge of the individual's birthchart. The second method relies on interpretation of the birthchart. The usual indications for individual use include the salt representing the sign which the sun was in at birth. We tend to use this salt in greater quantity and thus can find ourselves deficient if it is not replenished. The placement of Saturn in our chart gives an indication of things we tend to be lacking, so the salt for this sign is also used. In addition, the sign where the Moon's south node is placed indicates an area where we tend to "leak" energy, so this salt is also included. The Moon's south node is one of the two points at which the Moon's orbit crosses the ecliptic. These are held to be "karmic points", with the south node indicating our past and the things we cling to because they are so familiar.

Finally, we have what is called the "cell salt bridge". This is a combination of three salts we need the most and encompass the sun sign and two succeeding signs. So, for example, if one is born with the sun in Aquarius, they would take the salts related to Aquarius, Pisces, and Aries. The idea behind this comes from the fact that gestation takes nine months, during which time we receive each salt during its "season" provided by our mother. The experience of these salts and how to use them is passed on to us while we are in the womb. At birth, we find ourselves three months shy of a full circle of the signs and so we tend to be a bit uncoordinated in the use of these three missing salts.

The salts, being taken in homeopathic dilutions, do not so much replenish the bulk material but provide the body with clear information on how to utilize the individual salts.

Cell salts are also used in modern ayurveda. Their VPK factors are indicated in the following table.

VPK Factors of Cell Salts

Salt	Doshas
Potassium Phosphate	VP
Sodium Sulfate	P
Potassium Chloride	P
Calcium Fluoride	P
Magnesium Phosphate	VK
Potassium Sulfate	VP
Sodium Phosphate	V
Calcium Sulfate	P
Silica	P
Calcium Phosphate	VK
Sodium Chloride	V
Iron Phosphate	VPK

The cell salts are available in most health food stores in tablet form. Of course, these are not "philosophically" prepared materials and it would be ideal for the artist to prepare his own. The Acetate Path, which we will be examining in more detail later, provides a means for separating and purifying the Sulfur and Mercury of the minerals and metals. In a manner similar to creating plant stones, the mineral salt provides the Body which is reanimated with the parent metal's essence. For example, sodium chloride would be imbibed with the Sulfur and Mercury of sodium, then digested. The resulting matter would then be used as the starting material for homeopathic dilutions and subsequent use.

In either case, salts from the health food store or those prepared spagyrically can be enhanced in their effectiveness by taking them dissolved in the appropriate water fraction as we discussed earlier, in the anupana section.

Additional Attributions

In addition to the seven ancient metals and cell salts, there are many other materials used in alchemy, and each has its energetic associations. Some of the lesser-known planetary connections are presented in the following table.

Element	Planetary Ruler
Sodium	Jupiter
Potassium	Mercury/Uranus
Calcium	Saturn
Magnesium	Mercury
Aluminum	Moon
Zinc	Venus/Neptune
Platinum	Sun
Antimony	Earth/Uranus

The traditional correspondences for a selection of important mineral medicines used in ayurveda are shown below.

Stone	Planet	VPK Factor
Ruby	Sun	VK- P+
Garnet	Sun	VK- P+
Pearl	Moon	VPK=
Yel. Saphire	Jupiter	VP- K+
Yel. Topaz	Jupiter	VP- K+
Bl. Saphire	Saturn	VK- P+
Lapis Lazuli	Saturn	VP- K+
Emerald	Mercury	VP- K+
Diamond	Venus	P- VK+
Red Coral	Mars	P- VKo
Quartz	Venus	V- PKo
Onyx	Sun/Jup.	V- PKo
Amethyst	Saturn	VP- Ko

In the Chemical Sky of the alchemist, each planet (with the exception of Sun and Moon) occupies a day and a night house where it expresses positive or negative influence respectively.

This does not imply that one is good and the other is bad; instead, one is an active expression of the planetary energy while the other is passive.

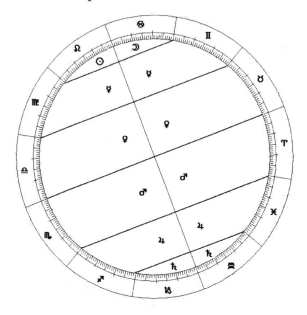

The energy of the system is said to enter in through the "Medium Coeli" or ceiling of the heavens, represented by the signs Capricorn and Aquarius. This relates to Saturn and Binah on the qabalistic Tree of Life, where the energy of the Absolute begins its decent into the material world.

CHAPTER ELEVEN

Alchemical Origin of Metals

In the alchemical tradition, minerals and metals have always been described as very powerful condensations of celestial energies, whose subsequent evolution towards perfection occurred in such immensely long cycles that they seemed to be immobile from our perspective. They are held to be in a state of deep sleep but their powers can be awakened.

As Basil Valentine instructs us in his work *The Triumphal Chariot of Antimony*:

> In very deed (that I may expound the matter in few Words) I found all Things, which are generated in the Bowels of Mountains, to be infused from the Superior Stars, and take their beginning from them, in the form of an aqueous Cloud, Fume or Vapour, which for a very long time fed and nourished by the Stars, is at length educed to a tangible form by the Elements.

You will find these same ideas concerning the generation of metals from a "celestial vapor" elaborated in many alchemical texts. Rudolf Glauber writes concerning this in the mid-1600s in his work titled *The Birth and Nativity of Metals*:

> The superior element Fire, as the Sun, Moon, and other Stars, send down their invisible virtues and fiery beams into the Earth's center, where they are congregated, and cause a great heat, and not being permitted to rest, leap back again and are scattered throughout the universal globe, and impregnate it with various and wonderful powers and are called minerals by the philosophers.

107

Every spiritual thing, come it from whatever body it will, is invisible and impalpable, nor can anything be made of it alone, but it's forced to remain a spirit, until it meets with a subject unto which it may adhere, be united, and by the benefit thereof be turned into a corporeal nature answerable to the purity of the subject and spirit, the spirit is in the room of seed; but the subject answers to the Earth or matrix in which the spirit is concocted into a sensible body suitable to its own nature.

When the Sun or any other star operates upon the moist Earth; the astral virtues are congregated and being made corporeal, do exhibit diverse minerals and metals according to the purity of the matrix, or moist Earth; where the water is instead of the matrix and the stars instead of the father, or seed: likewise, it is not possible for metals to be generated in the Center, where all things are dry, but far off from that place, where the waters moisten the Earth, and with which the Central Spirits can join themselves and pass into bodies and metals.

For a dry spirit cannot coagulate himself into a body by reason of his dryness, but wants a fit subject from whence to take its body, which is Water; as soon as ever the sulphureous spirit is mixed with the Water, it is no more common water, but the rudiment and beginning of a metallic generation called "Mercury" by the philosophers, not the vulgar, being already made metalline, but a viscous water, which the metallurgists call Gur or a Fermenting Spume, which if contained in a convenient place, and cherished with the due Central Heat and humidity, is in length of time matured into a metal.

108

The Conception therefore, and generation of metals is not in the profundity of the Earth by the mediation of the Central Spirits carried upwards, but also in the superficies by the stars casting their invisible beams into a subtle, and fat earth where they are held and become corporeal.

For the Sidereal Fire never ceaseth to infuse its virtues into the Earth, and to impregnate it with various products of vegetables, animals, and minerals, according as it meets with a matrix, nor is this done only in the Earth as being most fit for metallic generation, but even in the air in thick clouds do they act the same thing.

VPK Factors of Metals

Everything possesses a unique combination of the Five Elements which express through the Three Essentials of body, soul, and spirit. We described the so-called VPK Factors of various herbs and foods earlier. This provides us with a type of shorthand description of the expected effects of a substance and of the relative proportions of Mercury, Sulfur and Salt within it. Metals and minerals also have their VPK Factors.

Alchemists have described certain metals as having a greater proportion of Mercury or Salt than another which has more Sulfur. Each one is a unique combination of qualities which give it the properties we recognize as iron or lead or gold and so on. The following quote from Basil Valentine concerning the natures of metals illustrates the concept.

> Therefore the metal of Mars (Iron) is found to have the least portion of Mercury, but more of Sulphur and Salt.

> The reader must moreover know concerning the generation of copper, and observe that it is generated of much Sulphur, but its Mercury and Salt are in an equality.

> Among all metals Gold hath the pre-eminence because the sidereal and elementary operation hath digested and refined the Mercury in this Metal the more perfectly to a sufficient ripeness.

> Good Jupiter (Tin) possesses almost the middle or mean place between metals, it being not too hot, nor too cold, nor too dry, nor too moist, it hath no excess of Mercury, nor of Salt, and it hath the least of Sulphur in it....

> I tell thee that Saturn is generated of little Sulphur, little Salt, and much unripe gross Mercury, which Mercury is to be esteemed a froth that floats upon the Water in comparison of that Mercury which is found in Sol (Gold).

The alchemist Edward Kelly (1555-1597) presents a summary of the constituents we can expect to derive from the metals in his work titled *Theater of Terrestrial Astronomy*, written about 1580. Although he speaks mainly about the Sulfur and Mercury qualities acting through the body, his descriptions supplement Basil Valentine's with additional distinctions on the qualities of the Sulfur and Mercury. The terms Red and White are used to denote solar and lunar natures, respectively. Like Valentine above, Kelly provides a similar interpretation of the "VPK Factors" of metals.

The Sulfur/Mercury theory of metallic composition was popular prior to Paracelsus, who added Salt as the third component of the "Tria Prima", or Three Essentials present in all things.

> It is clearer than daylight that there are seven planets, seven days, seven metals, and seven operations. The metals are called after the planets, because of their influence and their mutual relations. The mineral principles are living Mercury and Sulphur. From these are generated all metals and minerals, of which there are many species, possessing diverse natures, according to the purity and impurity of the Mercury and Sulphur, resulting in the purity or impurity of the generated metal. Gold is a perfect body, of pure, clear, red Mercury, and pure, fixed, red, incombustible Sulphur. Silver is a pure body, nearly approaching perfection, of pure, clear, fixed white Mercury, and Sulphur of the same kind; it is a little wanting in fixation, colour, and weight. Tin is a pure, imperfect body, of pure, fixed and unfixed, clear, white Mercury outside, and red Mercury inside, with Sulphur of the same kind. Lead is an impure, imperfect body, of impure, unfixed, earthy, white, fetid Mercury and Sulphur outside, and red Mercury inside, with Sulphur of the same quality. Copper is an impure and imperfect body, of impure, unfixed, dirty, combustible, red Sulphur and Mercury. It is deficient in fixation, purity, and weight, while it

111

abounds in impure colour and combustible terrestreity. Iron is of impure, imperfect, excessively fixed, earthy, burning, white and red Sulphur and Mercury, is wanting in fusion, purity, and weight, abounding in fixed, impure Sulphur and combustible terrestreity. Nature transmutes the elements into Mercury, just as Sulphur transmutes the first matter. The nature of all metals must be the same, because their first substance is the same, and Nature cannot develop anything out of a substance that is not in it.

Everything nature produces is infused with a unique blend of the three essentials as a living entity. The alchemical art is all about assisting nature in the perfection of her creations. As Basil Valentine further teaches us:

For every Thing must have its own Matter; but not without Distinction. Animals require their Matter, Vegetables theirs, and Minerals theirs, yet from that formal Body must be extracted a certain Spiritual and Celestial Entity (shall I call it) or Apparency; for I find no other more fit name to give it: which Entity was by the Stars, before infused into that Body, and by the Elements concocted and made perfect. Yet this Spiritual Entity must again by a lesser Fire, and by the Regimen and Direction of the Microcosm, be reduced to a tangible, fixed, solid and incorruptible Matter.

Triumphal Chariot

Corrosion and Amelioration of metals

To the alchemist, the mineral realm is undergoing processes of evolution just as much as the animal and vegetable realms. All of the metals are on their way to becoming gold, the most noble metal, representative of the Sun and "the Sun behind the Sun". The other metals we are familiar with, such as iron, copper or lead, have been removed from their natural environment and smelted in order to utilize their specific properties, and in so doing, their further evolution within the Earth has been suspended.

Each of these metals was considered to be imperfect; they had defects (doshas) and thus were susceptible to their own form of disease, the cause of their corruptibility.

Eirenaeus Philalethes states it more clearly and concisely than most. He says:

> All metallic seed is the seed of gold; for gold is the intention of nature in regard to all metals. If the base metals are not gold, it is only through some accidental hindrance; they are all potentially gold.

Dhatus of the Mineral World

We talked about the seven dhatus or tissues that make up animal bodies; plants and minerals have their several types of tissue as well. The chart below indicates the dhatus and corresponding tissue along with its main function and corollary in the vegetable world.

Dhatu	Tissue	Function	Plant Part
Rasa	Plasma	Nutrition	Juice of Leaves
Rakta	Blood	Life Function	Sap/Resin
Mamsa	Muscle	Covering	Softwood
Meda	Adipose	Lubrication	Hard Sap/Gum
Asthi	Bone	Support	Bark
Majja	Marrow	Filling the space	Leaves
Shukra	Reproductive	Procreation	Flowers & Fruit

In ayurvedic literature, minerals and metals have an associated mythological origin. Most of the metals are referred to as semen of the gods and thus the most refined tissue of the mineral world that Nature produces. For example, mercury metal is called the "Semen of Shiva", gold is the "Semen of Vahni" (god of fire), and copper the "Semen of Kartikeya". Iron contains a Celestial Ambrosia, hidden by Shiva for the benefit of humanity. Alchemy takes up where Nature leaves off. Alchemical art refines and matures these "reproductive essences" into ojas and soma of the mineral world.

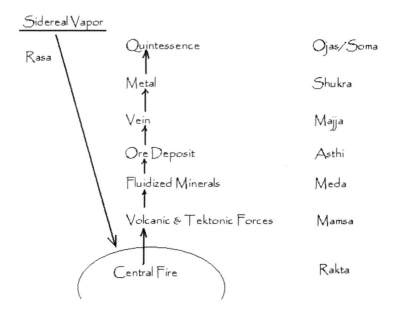

Many of the alchemical texts, both Eastern and Western, speak of the "amelioration" of metals; that is, making them better or curing them of their diseased condition for use as medicines or further elevating them to their proper end state, which is gold.

Through the Doctrine of Correspondences, the same laws and principles apply to all three kingdoms of nature, from the amelioration of plants as herbal elixirs to the amelioration of the "raw ores" of our daily life experience.

Metals have always provided the ultimate test in application of alchemical principles, and supplied valuable medicines to aid in accomplishing the Great Work of transforming the alchemist himself.

Outside of New Delhi in India, there stands the famous "Iron Pillar" a little over 20 feet in height and held by some to have been prepared by alchemical processes. Though it has stood exposed to the elements for about 1600 years, it shows no signs of corrosion.

CHAPTER TWELVE

Secrets of the Fire

In many of the ancient cultures, the soul of a thing or individual was represented by a flame standing alone or centered in the heart. In the Egyptian ritual of mummification, the heart was often removed and replaced by something more durable, a carved stone scarab, a solar symbol.

Ayurvedic tradition calls this fire "Agni"; we mentioned it earlier as the digestive fire. Agni has been translated as "the inner guide" and is described as the fire of spiritual consciousness, the "Divine Child", "The Seed", or "Embryo". It is the "Particular Fire" or Central Fire of an individual.

Agni, as the Divine Child, represents the soul or alchemical Sulfur which evolves through the kingdoms of nature. The soul is like a flame (Sulfur—"that which burns") that moves progressively through the mineral, plant and animal realms of nature; it is the fire that transforms their substance and evolves as the being that inhabits them.

The outer aspects of this fire include light, heat and color; the inner aspects include life, perception, and consciousness. Each one of us is a flame, a form of Agni, which is the fire that lights our body, breath, and consciousness.

Agni is said to be a portion of the solar fire which has descended into the Earth, which in turn the Sun has received from the light of pure consciousness, "the Sun behind the Sun".

The first level manifestation of this soul-fire is in the mineral realm.

The solar fire impregnates the Earth to its fiery core, as the alchemists describe in the last chapter, then rises upward as the mineral fire in volcanic and tektonic forces. The mineral fire builds up various crystals which are the first forms of life and powerful conductors of cosmic, spiritual, and physical forces. That is why the minerals and metals are so valuable as healing agents. Gold was held to be the purest form of Agni in the mineral world, like pure condensed sunlight.

The waters of the Earth are also impregnated by the solar fire and quickened by the fire of lightning, thereby raining down with an additional portion of this celestial fire.

The denizens of the mineral world are considered to be in a great "swoon", or deeply asleep from our perspective.

Cycles of rock and water eventually form soil, which sets the stage for the next manifestation of fire in the plant kingdom.

The soul that was locked in the mineral now awakens in the plant and learns to develop new forms of fire, particularly as photosynthesis, the digestive fire of plants, but also in the form of brilliant colors, spicyness, and fuels such as oils, wood, coal and petroleum; all rich storehouses of condensed solar power.

The plant is still intimately connected to the mineral soul which sustains it. The plant is unable to move or heat itself, so it requires the protection and heat the earth provides.

The fire works itself free from the earth as it moves into the animal world by developing yet more sophisticated modes of expression. The animal fire is built up around a digestive system (the first system to appear in the developing fetus) and various structures arise which allow it to hunt, capture, and assimilate fuel to keep the fire burning.

In addition to this digestive fire, the fire of respiration becomes much more active. Whereas the uptake of vital energy (prana) in the mineral and plant realms was passive, in the animal realm it becomes the active process of breathing, which can be regulated.

A plant exposed to constant smoke and dust can't move out of its path, but the animal can move to a better source of air loaded with a higher quality of prana.

These two fires, the earthy, food-based digestive fire and the solar derived prana or respiration-based fire, develop the animal body and mind. They are connected through the various circulatory systems (srotamsi), both physical and subtle.

The breath, carrying Prana or the solar fire, creates the mind and sensory organs. The earthy digestive fire creates the various tissues and motor organs of the body. In the lower animal world, the digestive fire predominates over the fire of

118

the breath, such that the motor organs, "the flesh", rule over the senses.

The human soul represents the height of evolution in the animal world. It evolves a new type of fire as reasoning intelligence. Humans still possess a refined animal or instinctual mind that can find food and protect the body, but also a higher intelligence that can digest ideas and discover the nature of reality beyond the body.

This fire of intelligence is a development out of the solar fire. It is fed by the essences of both the digestive fire and breath fire, the fixed and the volatile. In the human being, there is a battle between the solar fire of intelligence and the earthy fire of desire inherited from the animal kingdom. Our intelligence allows us to open up to the higher light of wisdom gained through experience, but when it is unbalanced, it often feeds the drives, passions and egoism of the lower nature. Through this solar fire, the human fire can transcend form and enter into higher states of awareness; it can free the flame of the soul imprisoned in dense matter.

The fire within us naturally seeks to return to its Divine solar home. Our soul's fire is the Divine Will within us seeking to return to the Absolute, the Celestial Fire. In this process of awakening intelligence, a new form of the fire, the so-called Kundalini or spiritual force, becomes active. Kundalini is a higher aspect of the Pranic fire. It rises like a flame from the base of the spine and carries our consciousness upwards out the top of the head and into the higher states of consciousness beyond the limitations of time and space.

The alchemists, the "Fire Philosophers", seek to consciously assist this transforming power of fire which propels us toward our ultimate perfection.

Similar to an oil lamp, our bodies (Salt) represent the wick with its many fibers, Prana (Mercury) is the fuel we extract from food, drink, and air, while Agni, the Sulfur or soul-fire, is the flame. With improper fuel or external conditions, some of the fibers in the wick can become clogged, making the flame burn unevenly. If it continues to become more clogged, the flame can flicker, burn low, or even be extinguished. The balanced flame burns clean and stable, and its brightness illuminates the darkness around us.

We can consciously select the fuels and conditions for our particular lamp to burn healthy and bright for a very, very long time.

In the laboratory we take care to nourish the fire inherent in our subject slowly and by degrees. You wouldn't throw a log onto smoldering tinder without first feeding it with small twigs, then larger and larger sticks until it is able to accommodate the log without being extinguished. So too in the laboratory our subject is gently matured until its soul fire can shine out in all its glory.

CHAPTER THIRTEEN

Introduction to Practical Works

The practical part of this book has been divided into three main sections in order to lead you step by step in the exploration of a few mineral works of alchemy with a few tips on plant alchemy thrown in for good measure. While we can't possibly cover all of the fascinating operations available to the student of the Art, I hope enough has been presented to firmly plant your feet on the path so you will feel confident to take up your own line of research with a solid background of both theory and practice.

The alchemical work on plants has been described in the highly recommended *Alchemist's Handbook,* by Frater Albertus, and *The Practical Handbook of Plant Alchemy*, by Manfred Junius (details in the bibliography). Little has been published which makes the mineral and metallic aspects of alchemy accessible to the modern student. I hope this present work will open that door.

The first section presents "The Acetate Path", a popular method of working with the mineral realm, steeped in ancient tradition. It provides an accessible method for preparing materials which can selectively act on the tissues and channels of the body to remove obstruction and restore balance to the elements. This path also provides the raw materials to create powerful menstruums which can unlock the entire mineral realm, and is one of the traditional Wet Ways of creating the Philosopher's Stone.

The second section is called "The Book of Antimony". Antimony holds a special place in Western alchemy as one of the great "Rasayanas". Held to have life-extending and rejuvenative powers, antimony is closely associated with metallic mercury, the main rasayana of Rasa Shastra. We will closely examine this fascinating mineral and preparations made from it, leading to its most valued essence, "The Fire Stone".

121

The work on antimony provides both a Wet and a Dry Way for confecting the Stone of the Wise.

The final section is "The Book of Gold", the incorruptible and immortal metal, congealed sunlight. We examine some of its history, and how to obtain it and work with it to prepare various alchemical medicines, including a few methods to prepare the famous "Aurum Potabile" or potable gold, the most sought after rasayana of the western world.

In each section, there are excerpts from the writings of alchemists which help to illustrate some of the practical methods applied to each material and process.

Laboratory Works

Before we start practical works, we should be familiar with some of the general guidelines, for safe operation. The practice of laboratory alchemy is not without a certain amount of risk and danger, so as they say, "safety is job one".

Safety and fire preparedness are an essential part of the alchemist's circle of protection in all works. Since fire is the main transformative tool in the laboratory, you will be working with various forms of it, so give it due respect and know how to subdue it if it gets out of control. A fire extinguisher is an essential piece of lab equipment.

A high school or college level chemistry lab class would be helpful for learning some of the basics of lab work. As an alternative, buy a kids' chemistry set (a good one) and work through the experiments given with it if you are an absolute beginner to lab work. This can provide you with an inexpensive hands-on training kit that will serve well for practical alchemical works later.

Know something about the materials you are working with and not just their names. Are they toxic? Flammable? Or generally safe? Such information can be found on the Internet or in such reference works as the Merck Index, CRC Handbook, or other chemical tables. There are Material Safety Data Sheets available online for most materials you are likely to encounter, at least as far as raw materials go. Alchemical products tend to be very complex and stretch the limits of identification at present, so don't expect to find an MSDS for the Oil of Gold.

There are many items of general cookware that make excellent labware (we will talk about these later); remember to keep them separate from your usual kitchen supplies. This will avoid any unpleasant contamination problems to your foods and your alchemical works; but also remember these are your tools of the Art and should be respected as such.

Preparation of Space

The practice of laboratory alchemy requires some space. If you have access to a room that can be dedicated to the purpose, that would be ideal; however, for most workers adequate space is at a premium and other options need to be explored. If you have room outside and zoning allows, one can often find an old camping trailer for little money or even for free. They make excellent labs which are isolated from the mainstream of daily life. The corner of a basement or garage which can be screened off is another option. I've seen spare closets turned into homey little lab spaces that were perfectly workable. Many artists don't have the luxury of space, but an old dresser in the corner of a room can provide a work surface and plenty of storage space between operations; others have an old trunk that doubles as a workbench and storage point.

Finally, some have no other option than a portion of the kitchen counter. There is a lot of activity in this area generally, so the possibility of breakage and contamination increases dramatically. If this is your only option, set up only what you need to do the work at hand and when finished disassemble and store your tools safely. Operating in a kitchen is marginally okay for plant works but not at all recommended for most mineral works.

Also, some operations need to be performed under adequate ventilation and unless you have a very good fume hood, this means you will need some space outside. A small yard or balcony space is fine, just have some consideration for your neighbors though; it is certain they will not like smelling noxious fumes coming from your calcining furnace. A trip out to uninhabited lands may be in order for certain operations if you live in close quarters to the neighbors. Check the laws in your area and keep it legal.

Once you have a space to work in, you should clean the area very well, and depending on your religious or philosophical preferences, perform a banishing or blessing to prepare the area for sacred works. Anything you can do to enhance the effect of entering sacred space when you undertake an operation will be helpful in all of your works. If you have space, a small altar or shrine within the lab is desirable.

Whether in the lab or outside of it, you should have a space which serves as a temple or shrine during meditations and charging of elixirs. You may have artwork in the lab as a constant reminder of the higher aspects of the Art and the fact that this is not just a chemistry experiment.

CHAPTER FOURTEEN

Guerrilla Alchemy

The practice of laboratory alchemy does not necessarily mean you will have to spend the proverbial arm and leg for equipment. In addition, the rise in illegal drug manufacturing and new antiterrorist laws have put a serious damper on the purchase of laboratory equipment and chemicals by the general public without coming under close scrutiny of law enforcement agencies. Fortunately there are alternatives we can take advantage of, many of which closely mirror the simple implements used by alchemists in times past. I tell my wife that alchemists have strange tastes in garbage. Trash to one is treasure to another.

Standard laboratory glassware is, of course, the ideal for all of these works; but it can be an expensive undertaking to equip your lab. It is also very disheartening to have your precious glassware reduced to rubble during some of the operations which place a heavy demand on equipment.

In many cases we can perform operations using the simple items you might find in any kitchen. With a little ingenuity we can accomplish our lab work in such a way that it will fit into anyone's budget. Thrift stores, yard sales, the local pottery supply house, and pet and aquarium suppliers are treasure houses for inexpensive labware.

Thrift stores always have assorted canning jars for extractions and storage. Also look for stainless steel flatware for calcining dishes, and stainless steel mixing bowls for sandbath or waterbath uses. Corningware type casseroles are also excellent for calcining or sublimations. They can take a red heat from a gas flame for hours without damage. There are many kitchen utensils that are useful in the lab, such as tongs, spatulas, coffee grinders and funnels. Study some of the old illustrations from alchemical texts that depict laboratory equipment and keep your eyes open for similar items of glass or ceramic. Various types of hot plates, slow cookers, water heating vessels and water pumps are also available at very reasonable prices. For heaters, try to obtain those that allow a continuous range of settings instead of the preset low, medium, and high range, as this will give a greater degree of control. As an alternative, a light dimmer switch, rated to the appropriate wattage, can be used to control heaters, pumps, etc.

Aquarium suppliers have glass and plastic tubing of various sizes as well as rubber and cork stoppers, valves and tubing clamps.

Pottery suppliers have an assortment of chemicals used for compounding glazes that are useful for alchemical work. Many of these are native minerals ground to a fine powder and include such materials as potassium carbonate or salt of tartar (also called "pearl ash"), niter, sal ammoniac, iron and copper oxides, antimony oxide, and vitriol.

You will also find crucibles, fire bricks and of course clay for sealing vessels in certain works or constructing your own apparatus.

Be sure to inspect the various pieces of glassware you plan to use for any evidence of flaws or cracks. Sometimes, glass knocking into glass or the sink while washing can produce small dings called Star cracks, because they look like little stars. In many cases these are still safe to use, but in heating or cooling rapidly, you run the risk of the glass breaking. It is very disappointing to see weeks or even months of work on a product lost because you didn't notice a small crack in the flask which ultimately lead to its failure. Practical alchemy teaches you to be aware of your surroundings at all times.

With these few hints on inexpensive labworking, let's take a look at how we can put it into practice, as we examine some of the most common operations.

Digestion

You can construct a simple chamber for digestion by placing a lightbulb inside a large plastic ice chest of the type used for camping. Simply by changing the wattage of the lamp you can control the temperature to the required degree. A simple indoor/outdoor thermometer is useful for monitoring temperature, which for general work ranges from about 90 to 104 degrees F. Many operators prefer to wrap vessels with aluminum foil to protect them from the bright light.

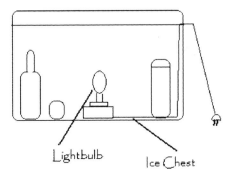

Lightbulb   Ice Chest

Simple Digester

As an alternative, you can pour in a layer of sand at the bottom of the ice chest and place a heating pad on top of it. Pour another layer of sand 2 to 3 inches deep on top of the heating pad. Now your flasks and containers can be nestled down into the sand for a gentle surrounding warmth. With a little luck, you can get up to a year's worth of service from the heating pad before it needs replacing.

If water for cooling is problematical, you can use a gallon jug suspended several feet above your apparatus, then use a siphoning tube to feed condensers, etc., with a second jug as a drain. A good alternative to this is to use a small submersible water pump such as those used in tabletop fountains. Place the pump into a large bowl as a water reservoir and use it to feed into condensers, etc., then run the drain line back into the bowl so you have a continuous loop of cooling water.

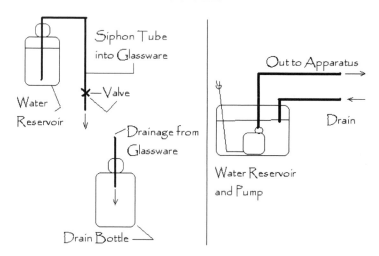

Extraction

Extraction refers to methods for separating the pure from the impure. As with all of these basic operations, there are many ways to accomplish the end.

In general, extraction methods use an extraction medium to effect separation. When you brew tea, water is the extraction medium; when you made the seven herbal basics, vodka was the extraction medium. Remember that the Mercury of a kingdom has an affinity for the Sulfur of that kingdom. In herbal extraction we use alcohol, the vehicle for vegetable Mercury, to pull out the Sulfur of the plant. The resulting extract is called a tincture, from the Latin *tinctura*, that which colors. It contains the Mercury and Sulfur of the plant. The extracted plant residue contains the salt which we obtain by calcination.

There are three methods widely used to effect extraction of a tincture from plants or minerals; the first is called Maceration. We used this method to make the seven basics.

Pour the extraction media, also called the solvent, over the material to be extracted until it stands about two finger widths above the material. Seal tightly and shake well. Place into a warm spot to digest, and remember to shake it daily. After the extraction period, which can be a few hours to over a year in some operations, simply filter the tincture to separate it from the residue.

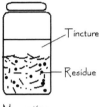

Maceration

The second type of extraction allows us to use a greater heat without loss of the volatile components. This method is called a Reflux Extraction. The matter to be extracted is placed into a flask and covered with the extraction medium. Then a condenser is attached to the top and heat is applied such that condensation occurs no higher than the first 1/3 of the condenser. After several days of this constant heating without liquid loss, the material is allowed to cool, then filtered to obtain the extract and solid residue.

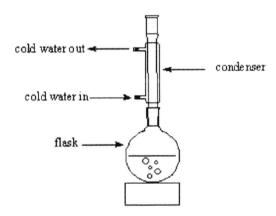

If you do not have a condenser, a two- or three-foot long glass tube will often suffice, and if the temperature is carefully regulated, the ambient air will provide cooling.

The third method of extraction is called Soxhlet Extraction, named after Franz Von Soxhlet, a German chemist who invented a special glassware apparatus for extracting fatty substances from materials in 1879. The apparatus has come to be called a Soxhlet Extractor and can be somewhat expensive. In this method, the material to be extracted is placed into a filter paper cup, called a "thimble", and inserted into the extractor body. The extraction medium is placed into the flask at the bottom and heat is applied to it. The solvent vapors rise up a side tube into a condenser where it turns to liquid again and drops into the extractor body and thimble. The extractor body fills with solvent until it reaches the top of the siphon tube, whereupon it drains back into the flask. This cycle repeats until extraction is complete and there is no more coloring of solvent in the extractor body. The extract or tincture is recovered from the flask and the extracted residue is recovered from the thimble. Very often several thimbles are filled and extracted with the same solvent to obtain a concentrated tincture.

Condenser

Extractor body with side tube and siphon tube. The Thimble containing material to be extracted is placed inside.

Flask containing solvent

A soxhlet extractor is a nice investment but not essential; it is definitely not something you will find in a thrift store.

Distillation

Distillation is an important process in alchemy and is sometimes a stumbling block to those without standard chemistry glassware. There are simple ways to perform this using common items and though not as elegant as "official" glassware, they get the job done in many cases.

At the very simplest, a jar containing liquid to be distilled can be placed into a larger jar which is then sealed and placed in a sunny spot. The distillate will form on the walls of the large jar and run down the sides to the bottom where it is collected. It may not be terribly efficient or suitable for all distillations, but it is easily available for anyone to experiment with. There are many illustrations in old texts depicting this type of solar distillation.

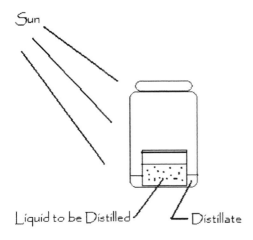

Okay, maybe there is an even simpler method. It may seem silly but it actually works; I call it "Distillation by Baggie". Just place a small amount of material to be distilled into a dark colored dish and set it into a large re-sealable plastic freezer bag. Seal the bag and place it onto a windowsill that gets a lot of sunlight all day. Trail the empty part of the bag over the edge of the sill so it is in the shade. As the sun heats the liquid in the dish, condensation will form, then drip down to the bottom. To collect the distillate, simply snip off the bottom corner of the bag and let it drain into a receptacle.

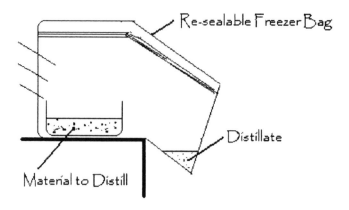

The simple distillation train shown below uses a stainless steel teapot connected to a half-gallon glass jug with cork stoppers and a length of ¼ inch stainless steel tubing. Remember to always allow a vent hole to relieve pressure at the receiver end. A little duct tape around the teapot lid helps reduce losses that could occur if it is not tight. While not very elegant, it works just fine to collect essential oils and costs very little.

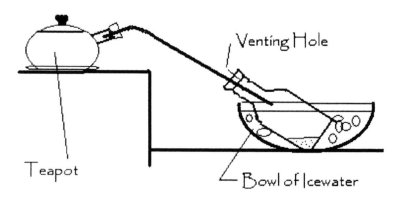

Of course in more advanced operations, there is no substitute for chemistry glassware. You can find good deals online, such as on eBay, for glassware kits with multiple pieces of standard taper items. These can be fitted to perform a multitude of operations.

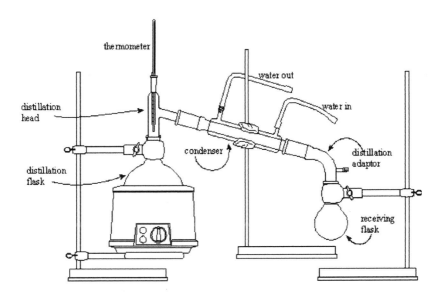

The diagram above shows the various parts necessary for a distillation train using standard chemistry glassware. Notice the distillation adapter connected to the receiver provides a small tube as a vent or connection to vacuum.

It is often necessary to keep moisture out of the system and yet maintain its ability to vent excess pressure. This can easily be accomplished by attaching a balloon to the vent outlet tube or by attaching a simple drying tube as shown below.

The drying tube is a short open piece of tubing filled with a drying agent like silica gel or potassium carbonate which will allow the system to vent but keep atmospheric moisture from getting in. A six-inch piece of half-inch inside diameter PVC pipe will work for this.

The addition of a few small "boiling stones" to the distillation flask is also recommended. These are small pieces of pumice, about the size of rice, which stabilize the boiling liquid and prevent sudden ebullitions from local hot spots in the flask.

A simple high temperature retort can be constructed using iron or stainless steel pipe fittings as shown below.

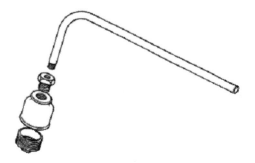

This type of retort is an old miners' trick, used to distill mercury amalgams containing gold from panning. A fireproof pan is filled with charcoal and the retort is placed into the center with the distillation tube immersed into a container of water which condenses the mercury vapors.

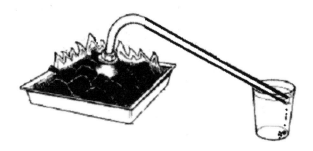

While it can also be used to distill antimony regulus amalgams, which we will talk about later, it is not recommended for very corrosive materials.

Be sure to withdraw the tube from the water before allowing the retort to cool, as a vacuum will be created that may pull cold water in and burst the hot retort.

Vacuums

Distillations are often performed under a vacuum. This allows the operation to proceed using much less heat and thus prevents scorching delicate oils and aids preserving the life in them.

In the old days, alchemists would heat their flasks full of material, then quickly seal on a heated receiver. On cooling, the apparatus had a vacuum inside.

We have an advantage today with vacuum pumps and accurate gages to apply and monitor a system under vacuum. Mechanical vacuum pumps can be expensive. As an alternative, the pump and motor from an old refrigerator can be used to provide adequate vacuums. There are also small handheld vacuum pumps and water aspirator pumps available that work quite well. Once the vacuum is applied, you will need a way of seeing that it is holding.

An inexpensive alternative to vacuum gages from the chemistry supply house is a simple automotive vacuum gage from the auto parts store. Before applying heat to your apparatus, it is a good idea to place the vacuum on it and observe it for an hour to see if it leaks. Silicone lubricants are used to help form a good seal, but they should be applied sparingly to avoid contaminating your products.

A word of caution: if your apparatus is closed and under vacuum, do not reapply the vacuum to the heated system if the vacuum slowly leaks out during an operation. This could cause a sudden rush of liquid streaming over into the receiver and you will have to start over. It is safer to let the apparatus cool, then reapply the vacuum before continuing.

Sublimation

There are some solids you will come upon in alchemy which are purified by sublimation. These materials go from the solid state to a vapor without passing through a liquid state, and the vapor returns to the solid state upon contact with a cool surface. Alchemically, the body opens and the finer parts ascend; these are captured by a cool condensing surface in an "exalted" form. Most notable are many of the ammonia-based salts such as Sal Ammoniac, but there are also many metallic

135

compounds which are purified by sublimation such as those from antimony, mercury and zinc.

CorningWare casseroles work well for this. Place the matter to be sublimated in a layer on the bottom of one casserole, then cover with a second casserole that is inverted. A strip of clay-smeared cloth can be used to seal the joint and dried before use. Gently heat the bottom and the sublimate will collect on the upper surfaces.

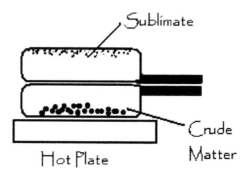

The final sublimation temperature will depend on the matter you are sublimating and can range from near room temperature to a full red heat.

Filtration tips

The separation of solids from a liquid by filtration is a very common procedure in alchemical works. In general, paper filters serve most purposes and come in a wide variety of grades. Coffee filters can be used in the majority of works, especially for the work on herbs.

Filter papers from a chemical supply house or the local winemaking shop are usually of a finer type and can handle most filtration problems on the mineral or herbal work. Some materials are too acidic or alkaline for papers to survive very long. In such cases, a small wad of glass wool can be used in place of a paper filter.

You can use plastic 2 liter soda bottles for funnels by cutting off the top 1/3. This top piece will easily hold most sizes of coffee filter.

Coffee
Filter

Top of
Soda Bottle

Nylon stockings make wonderful filters for many materials. They can hold a tremendous amount of solids and can be squeezed quite hard to remove residual liquid without breaking as paper filters do. They are especially useful in herbal works to remove the bulk of the herb residue before a final filtration through paper. However, they do not stand up to acidic conditions or concentrated, hot alkalis.

Don't overdo filtering as there is loss of product each time.

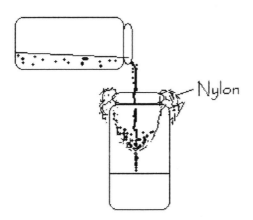

Nylon

The term distillation, as a means of purifying a liquid, was often used to describe filtration as well in old texts. Thus you may come upon the phrase "Distillation by Filtration" as part of a process. Before the general use of filter papers, the ancient artists used wicks of cotton or wool to filter a liquid from container to container as illustrated below. This technique is useful when the liquid you wish to filter refuses to pass through a paper or has a large amount of very fine suspended, gelatinous material.

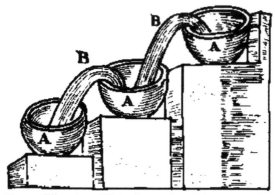

From *The Art of Distillation*, by John French, 1651

Metal cans

Let's not forget the lowly tin can. We throw them out by the dozens, but they have many uses in the laboratory. They make strong supports for equipment or to raise heaters off the tabletop to prevent scorching.

You can use them to incinerate herbs in if they are first heated to remove any plastic interior coating.

A simple outdoor hot plate which is heated by wood or charcoal is useful for evaporating malodorous liquids or toxic materials, away from inhabited areas.

Coffee Can Hot Plate

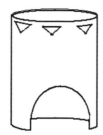

You can construct a simple furnace by nesting one can inside of another and then pouring in a layer of insulating material such as vermiculite, pearlite, or even sand or kitty litter.

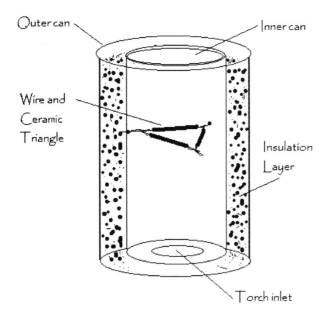

Outer can

Inner can

Wire and
Ceramic
Triangle

Insulation
Layer

Torch inlet

Cut a hole in the bottom to insert a small propane torch. Heavy wire can be poked into the inside can to form a triangular support for crucibles. This furnace will get hot enough to make regulus or glass of antimony (more on this later). Although the furnace has a limited lifespan, it costs practically nothing.

For a more durable installation, you can make your own castable refractory cement for building furnaces. Obtain a container of high temperature furnace mortar from the hardware store and a bag of pearlite soil conditioner from the garden supply. Mix about one part of the mortar with four parts pearlite to form a sticky mass. Mold the material into bricks or whatever shape needed to create your furnace hot box, then let it dry completely before use.

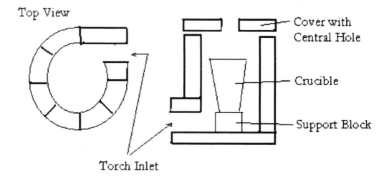

Top View

Cover with Central Hole

Crucible

Support Block

Torch Inlet

Micro methods

A microscope is a very useful tool in the laboratory—not a superpowerful one, but a magnifier of 30X to 60X. With a wide field microscope, you will be able to characterize materials in various ways. You will also be able to follow reactions on a small scale before you proceed with a larger preparation using precious materials that have taken months to purify.

A small adjustable hot plate mounted below the microscope can provide a "hot stage" on which you can observe melting points, sublimations, distillations, in fact any of the laboratory operations, using 100 mg or less of total material. This can lead to a tremendous conservation of materials. You may come to a point in an operation where it is unclear how to proceed, or there are several options to explore. You can often make very small tests with portions of the matter until it is clear.

Using micro methods is especially useful for experimenting with very toxic materials and rare materials.

Examination of solids or liquids under polarized light, laser light, or ultraviolet lighting can add another dimension to alchemical changes in your matter.

Chemical Analysis

Appendix II presents a small selection of analytical data collected on some of the alchemical products we will be examining. This data has been collected using state of the art chemical instrumentation including gas chromatography, mass spectrometry, infrared spectrometry, and plasma emission spectroscopy. They are not detailed analytical reports, but serve only to indicate the types of substances obtained and their complexity. This data will be of use to future researchers involved with comparing and identifying the physical components of alchemical products, though they give no indication of the subtle energies associated with them. For example, in plant alchemy, the Mercury of the plant is said to be carried by the alcohol. We can identify alcohol, the body, but with the present technology we can make no determination as to the nature or quality of the subtle principle which is the alchemical Mercury. Modern instrumentation has limitations in the detection of "subtle forces" of alchemical works, similar to modern medicine's inability to accurately map acupuncture meridians.

Our senses, taste, smell, sight, sound, and feel, were the analytical instruments of the ancients. Their assessment of a substance was coupled with developed psychic and intuitive perceptions of the material's inner nature.

Modern chemical instruments stretch the limits of our senses into areas we don't normally perceive. For example, when materials are intensely heated, they emit characteristic frequencies of light. Spectral analyzers can qualitatively and quantitatively determine these characteristics precisely and in areas of the spectrum we can't even see. The visible spectrum, extending through ultraviolet and x-rays, provides identity of elements. At the other end of the spectrum, infrared and microwaves, we can see the bonding between elements and how that bond is stretching, bending, or bouncing back and forth. From this we can deduce a structure and identity for the material body, but at present we have no reliable method for determining the quality of its inherent intelligence or life force. One day, future instrumentation will bridge this gap.

CHAPTER FIFTEEN

The Acetate Path

The so-called acetate path is a "Wet Way" for uniting vegetable nature and metallic nature by infusing plant life into a mineral or metallic body in order to accelerate its evolution. Plant life still has a strong affinity for the mineral realm and provides a medium through which the mineral powers are awakened, purified and made acceptable to the human body.

History

Man's association with mineral and metallic acetates dates back several thousand years, at least to the early arts of winemaking and the production of vinegar.

The action of vinegar (which is mainly a dilute acetic acid) on many mineral or metallic bodies of ancient use produces what we today call the metal acetates. Prior to the 1700s, chemical terminology was not very well defined, so you will often read alchemical texts wherein an acetate of a mineral or metal is referred to as a vitriol because they are generally glassy in appearance and water soluble, just as the more commonly recognized copper and iron vitriol (ie, copper sulfate and iron sulfate).

The acetates from the mineral realm form the starting material for an interesting and valuable method of obtaining the Three Essentials of this kingdom.

Many of the alchemical adepts describe the process of using acetates, or their end products, in varying degrees of subtlety and veiled terminology.

The following excerpts in the alchemist's own words illustrate the method and some of its amazing potential.

143

Ripley

One of the earliest clear descriptions of the process comes to us from Sir George Ripley, the English alchemist who lived from about 1415 to 1490. The following text on his work was published much later and describes one of the methods for confecting the Philosopher's Stone. It gives an indication of just how far the "Acetate Path" can take us. This excerpt describes only the first portion of creating the Stone. In the beginning of the text, Ripley speaks of taking Sericon or Antimony as the matter to be worked upon. This is usually translated as meaning lead oxide, but for our purposes in using the method, we can translate this to be any of the metals. We will come back to the importance of lead to this method later.

> The Bosome-Book of Sir George Ripley, Canon of Bridlington. Containing His Philosophical Accurtations in the makeing the Philosophers Mercury and Elixirs.

> London, Printed for William Cooper, at the Pelican in Little Britain. 1683.

> **The whole Work of the Composition of the Philosophical Stone, of the great Elixir, and of the first Solution of the gross Body.**

> First take 30 pound weight of Sericon or Antimony, which will make 21 pound weight of Gum, or near thereabouts, if it be well dissolved, and the Vinegar very good, and dissolve each pound thereof in a Gallon of twice distilled Vinegar when cold again, and as it standeth in Dissolution in a fit Glass Vessel, stir it about with a clean Stick very often every day, the oftner the better, and when it is well molten to the bottom, then filter over the said Liquors three several times, which keep close covered, and cast away the Feces, for that is superfluous filth which must

144

be removed, and entreth not into the Work but is called Terra damnata.

The making of our Gum or Green Lyon.

Then put all these cold Liquors thus filtered into a fit Glass Vessel, and set it into Balneo Mariae to evaporate in a temperate heat, which done our Sericon will be coagulated into a green Gum called our green Lyon, which Gum dry well, yet beware thou burn not his Flowers nor destroy his greeness.

The Extraction of our Menstue, or Blood of our Green Lyon.

Then take out the said Gum, and put it into a strong Retort of Glass very well Luted, and place it in your Furnace, and under that at the first, make sober Fire, and anon you shall see a faint Water issue forth, let it waste away; but when you see a white Smoak or fume issue forth, then put too a Receiver of Glass, which must have a very large Belly, and the mouth no wider then it may well receive into that the Neck of the Retort, which close well together that no fume issue forth of the Receiver. Then encrease your Fire by little and little till the fume which issueth be reddish, then continue the greater Fire, until drops like blood come forth, and no more fume will issue forth, and when that leaveth bleeding let it cool or asswage the Fire by little and little, and when all things are cold, then take away the Receiver, and close it fast suddenly, that the Spirits vanish not away, for this Liquor is called, our Blessed Liquor, which Liquor keep close stopped in a Glass till hereafter. Then look into the Neck of the Retort, and therein you shall find a white hard Ryme as it were the Congelation of a Frosty vapour or much like sublimate, which gather with diligence and keep it apart, for therein are contained great

Secrets which shall be shewed hereafter, after the great Work is ended.

The Creation of our Basis.

Then take out all the Feces which remaineth in the Retort, and are blackish like unto Soot, which Feces are called our Dragon, of which feces Calcyne one pound or more at your pleasure in a fervent hot Fire in a Potters or Glass-makers Furnace, or in a Furnace of vente (or a Wind Furnace) until it become a white Calx, as white as Snow, which white Calx keep well, and clean by it self, for it is called the Basis and Foundation of the Work, and it is now called Mars, and our white fixed Earth or Ferrum Philosophorum.

The Calcination of the Black Feces called our Black Dragon.

Then take all the rest of the aforesaid black feces or Black Dragon, and spread them somewhat thin upon a clean Marble, or other fit Stone, and put into the one side thereof a burning Coal, and the Fire will glide through the Feces within half an Hour, and Calcyne them into a Citrine Colour, very glorious to behold.

The Solution of the said Feces.

Then dissolve those Citrine Feces in such distilled Vinegar, as you did before, and then filter it likewise, three times as before, and after make or evaporate it to a Gum again, and then draw out of it more of our Menstruum, called now, Dragons Blood, and iterate this Work in all points as afore, until you have either brought all, or the most part of the Feces into our Natural and Blessed Liquor, all which Liquor put to the first Liquor or Menstrue called the Green Lyons Blood, and set that Liquor then altogether in one Vessel of Glass fourteen days in Putrification, and after proceed to the Separation of Elements, for now have you all the Fire of the Stone in this our Blessed Liquor, which before lay hidden in the Feces, which Secret all the Philosophers do marvellously hide.

Isaac Holland

Isaac Holland describes the work on Lead Acetate, in *Opus Saturni* or *A Work of Saturn*.

My child shall know, that the Stone called the Philosopher's Stone, comes out of Saturn. And therefore when it is perfected, it makes projection as well in man's Body from all Diseases, which may assault them either within or without, be they what they will, or called by what name soever, as also in the imperfect Metals.

And know, my Child, for a Truth, that in the whole vegetable work there is no higher nor greater Secret than in Saturn; for we do not find that perfection in Gold which is in Saturn; for internally it is good Gold, herein all Philosophers agree, and it wants nothing else, but that first you remove what is superfluous in it, that is, it's impurity, and make it clean, and then that you turn it's inside outwards, which is it's redness, then will it be good Gold; for Gold cannot be made so easily, as you can of Saturn, for Saturn is easily dissolved and congealed, and it's Mercury may be easily extracted, and this Mercury which is extracted from Saturn, being purified and sublimed, as Mercury is usually sublimed, I tell thee, my Child that the same Mercury is as good as the Mercury which is extracted out of Gold, in all operations; for if Saturn be Gold internally as in truth it is, then must it's Mercury be as good as the Mercury of Gold, therefore I tell you that Saturn is better in our work than Gold; for if you should extract the Mercury out of Gold, it would require a year's space to open the body of Gold, before you can extract the Mercury out of the Gold, and you may extract the Mercury out of Saturn in 14 days, both being alike good.

After preparing and purifying lead acetate crystals, he continues,

.... take the Matter out, and put it into a thick glass which can endure the Fire, set a head on it, put it in a Cupel with Ashes, which set on a Furnace, first make a small Fire, and so continually a little stronger, till your Matter come over as red as Blood, thick as Oil, and sweet as Sugar, with a Celestial Scent.

Reiterate this distillation in the Bath until the Matter hath no more Spirit of the Vinegar in it, then take it out, set it in a glass pot, distill all that will distill forth in ashes, till the Matter become a red Oil, then have you the most noble Water of Paradise, to pour upon all fixed stones, to perfect the Stone; this is one way. This water of Paradise thus distilled, the Ancients called their sharp, clear Vinegar, for they conceal its' name.

Paracelsus

The writings of Paracelsus often mention the use of spirit of wine, but the reader is left to decide whether he means alcohol from wine or Philosophical Spirit of Wine obtained by acetate distillation.

In his work titled *A Hundred and Fourteen Experiments and Cures*, Paracelsus has this to say about the work on lead and specifically on the "Stone of Saturn":.

The philosopher's Saturn is properly the Markasite of Lead, and in deed doth excel Sol and Luna, in so much that Raymond saith, that in this inferior world, there is no greater secret than that which consisteth in the Markasite of Lead...

The markasite of lead is lead sulfide as it occurs in the mineral Galena. Paracelsus directs one to extract this ore with vinegar; evaporate the vinegar to obtain a solid which is distilled as described in the excerpts above. He calls the clear distillate "Aqua Philosophorum" or "Ardantem" and also mentions the red oil.

The residue, or Black Dragon, is to be calcined and extracted with vinegar again. The vinegar is evaporated off and the residue calcined to redness a second time. This red calx is dissolved in vinegar, then evaporated to obtain fine clear crystals which are the prepared Salt.

This Salt is ground fine and saturated with the red oil, then digested in a sealed vessel for 21 days, after which it will be fixed into a red stone. The red stone is then ground to a powder and fine gold calx is added as a ferment. Add to this mixture an equal weight of the Aqua Ardens or Ardantum, seal the vessel and digest in a gentle heat until it is fixed. During this digestion various colors will appear, and in the end you will have "The True Oil and Elixir, ... which being duly prepared, doth not only alter and change the filthy and corrupt humors of our bodies, but also can change and transmute Luna into Sol".

Weidenfeld/ Spiritus Vini Philosophici Lulliani

In 1685, Johannes Segerus Weidenfeld published his work called *Secrets of the Adepts*. The main premise of the work was to show that there was a central ingredient necessary for the special menstruums used by the alchemical adepts and hidden by a variety of names. He calls this "Spiritus Vini Philosophici" or The Secret Wine Spirit of the Adepts.

He cites many alchemical texts extensively, including those of Raymond Lull, Paracelsus, Isaac Holland, George Ripley, and Basil Valentine, in order to show that what they term "Spirit of Wine" used in the preparation of their various menstruums, alkahests, and circulatums is not common spirit distilled from wine but in fact the products derived from the acetate process.

The following excerpts from his work illustrate his argument.

> I found an easie way from Paracelsus to Lully, Basilius, and other Philosophers of the same Faculty, who I saw agreed all unanimously in confirmation of the Paracelsian Menstruums; yea Light adding Light to Light, appeared so clear, that their preparation, variety, simple and literal sense showed themselves all at once, one only word remaining unknown, yet expressing the universal Basis of all the Adepts, and that is Spirit of Wine, not Common, but Philosophical; which being known and obtained, the greatest Philosophical, Medicinal, Alchymical, and Magical Mysteries of the more secret Chymy, will be in the power of the Possessor.

> You will find, that all the Secrets of Chymy depend upon one only Centre of the Art, namely the Spirit of Philosophical Wine.

> But by Menstruum we mean a volatile Liquor made several ways of the Spirit of Philosophical Wine and divers things.

Weidenfeld describes the preparation of twenty-four classes of menstruums effective in the vegetable, animal, and mineral worlds, all of which are centered around the use of the Philosophical Spirit of Wine which is "sharpened" by processing it with various herbs, salts, acids and alkalis. Weidenfeld uses the term "acuating" from the Latin *acuere*, meaning to sharpen or needle-like.

He also calls it "Vegetable Mercury" because it has the power to preserve the generative principles of the material being extracted.

The twenty four following Kinds of Menstruums will prove, that amongst the Dissolvents of the Adepts, no one is made without the Vegetable Mercury, or Spirit of Philosophical Wine: for it is the foundation, beginning and end of them all.

It is necessary to observe that the Spirit of Philosophical Wine appears in two forms; either like an Oyl swimming upon all Liquors, or like the Spirit of Common Wine.

The latter hath these more general Names in the Latine Tongue, Essentia Vini, Alcool Vini, Mercurius Vini, Vinum Vitæ, Vinum Salutis, Aqua Vitæ, Aqua ardens, Vinum adustum, Vinum sublimatum, Anima, and Spiritus Vini.

The method which we used in the Vegetable Menstruums, we will as near as we can observe also in these Mineral Menstruums: In the Vegetable we extracted from the Philosophical Wine an Aqua ardens, from which we did by Circulation separate an Oyl or Essence of Wine, which is our Spirit of Wine, which then by acuating divers ways we reduced into the precedent Kinds of Vegetable Menstruums, but in the Mineral we will begin with Philosophical Grapes, the Matter itself of Philosophical Wine, which is elsewhere called Green Lyon, Adrop, etc.

The process of distilling the metal acetate in order to prepare the philosophical wine is clearly described as Weidenfeld quotes Ripley:

> ... making at first a gentle Fire, but not luting the Receiver, till the Phlegm be gone over, and this continue, till you see fumes appear in the Receiver white as Milk; then increasing the Fire change the Receiver, stopping it well, that it may not

evaporate, and to continually augment the Fire, and you will have an Oyl most red as Blood, which is airy Gold, the Menstruum foetens, the Philosophers Sol, our Tincture, Aqua ardens, the Blood of the Green Lyon, our unctuous Humor, which is the last consolation of Man's Body in this Life, the Philosophers Mercury, Aqua solutiva, which dissolves Gold with the preservation of its Species, and it hath a great many other Names.

By Green Lyon, the Philosophers mean the Sun, which by its attractive Virtue makes things Green, and governs the whole World.

The Green Lyon is that, by which all things became Green, and grow out of the Bowels of the Earth by its attractive Virtue, elevated out of the Winter Caverns, whose Son is most acceptable to us, and sufficient for all the Elixirs, which are to be made of it; for from it may be had the power of the white and red Sulphur not burning, which is the best thing.

These Menstruums they called White Fume, because of their white and opake Colour. It is also called White Fume, saith Ripley, not without cause, for in distillation a white fume goeth out first, before the red Tincture, which ascending into the Alembick, makes the Glass white as Milk, from whence it is also called Lac Virginis, or Virgins Milk. In the same place: Out of the red Fume or red Tincture, otherwise call'd the Blood of the Green Lyon, the Adepts did by rectification alone prepare two Mercuries, namely, red and white: Upon this occasion, saith Ripley, I will teach you a general Rule: If you would make the white Elixir, you must of necessity divide your Tincture (the Blood of the Green Lyon) into two parts, whereof one must be kept for the red Work, but the other distill'd with a gentle Fire; and you will obtain a white Water, which is our white Tincture, our Eagle, our Mercury and Virgins Milk: When you have these two Tinctures, or the white and red Mercury, you will be able to practise upon their own Earth, or upon the Calx of Metals; for the Philosophers say, we need not care what substance the Earth is of, etc.

Ripley the first, and indeed the only man of all, declares to us, that the Key of all the more secret Chymy lyes in the Milk and Blood of the Green Lyon, that is, that the stinking Menstruum (or the parts of it, Mercury and Sulphur, Virgins Milk and the Lyons Blood, white and red Mercury) being fourteen Days digested gently, is the white and red Wine of Lully, and other Adepts.

Concerning these three substances of the stinking Menstruum, Ripley hath these following Sayings, in his Book named Terra Terræ Philosoph., where thus: When therefore you have extracted all the Mercury out of the Gum, know, that in this Mercury are contained three Liquors, whereof the first is a burning Aqua vitæ, which is being extracted by a most temperate Balneo: This Water being kindled, flames immediately, as common Aqua vitæ, and is called our attractive Mercury, with which is made a Cristalline Earth, with all Metallick Calxes also. After that there follows another Water thick and white as Milk, in a small quantity, which is the Sperm of our Stone, sought by many men; for the Sperm is the Original of men and all living Creatures; whereupon we do not undeservedly call it our Mercury, because it is found in all things and all places; for without it no man whatsoever lives and therefore it is said to be in every thing. This Liquor, which now you ought to esteem most dear, is that Mercury, which we call Vegetable, Mineral, and Animal, our Argent vive, and Virgins-milk, and our permanent Water: With this Mercurial Water we wash away the Original sin, and pollution of our Earth, till it becomes white, as Gum, soon flowing; but after the distillation of this aforesaid Water, will appear an Oyl by a strong Fire; with this Oyl we take a red Gum, which is our Tincture, and our Sulphur Vive, which is otherwise called the Soul of Saturn, and Living Gold, our precious Tincture, and our most beloved Gold.

The Virgins milk, or white Mercury (otherwise the white Wine of Lully) extracted out of the Green Lyon is by Paracelsus that Glue of the Eagle, or Green Lyon, so carefully sought for: For Eagle and Green Lyon are to the Adepts Synonymas of the same thing: For thus Ripley before: You will obtain the white Water, which is our white Tincture, our Eagle, our Mercury and Virgins-milk. Consequently therefore, red Mercury (or the red Wine of Lully) is the Blood of the Red or Green Lyon: For the same Lyon is called sometimes Green (in his Youthful Estate) sometimes red (in his more grown Estate) and therefore the Blood is sometimes said to be of the Green Lyon, sometimes of the Red.

That the milky Liquor or Spirit, Virgins Milk, white Mercury, the White Wine of Lully, and the Glew of the Green Lyon, called by Paracelsus the Glew of the Eagle, are terms synonymous; and that the Red Liquor, Blood of the Green Lyon, Red Mercury, the Philosophers Sulphur, and the Red Wine of Lully, otherwise by Paracelsus, the Blood of the Red Lyon, are likewise Synonymas.

Then will ascend a white Fume, which will make
the Vessel look like Milk, which must be gathered,
till it ceaseth, and the Vessel is returned to its
former colour: For that Water is the Stinking
Menstruum, wherein is our Quintessence, that is,
the white Fume, which is called the Fire against
Nature, without which our Natural Fire could not
subsist, whereof we will say more in its proper
place: And these, namely, the Mineral and
Vegetable Water, being mix'd together, and made
one Water, do operate contraries, which is a thing
to be admired; for this one dissolves and congeals,
moisteneth and dryeth, putrefies and purifies,
dissipates and joyns, separates and compounds,
mortifies and vivifies, destroyeth and restoreth,
attenuates and inspissates, makes black and white,
burneth and cooleth, beings and ends. These are
the two Dragons fighting in the Gulf of Sathalia,
this is the white and red Fume, whereof one will
devour the other.

The Spirit of our Wine, is the Basis, Root, and
Center of all Menstruums, Medicines, Alchymical
Tinctures, and Precious Stones,
Without these Waters we do little Good in this
Art; but he that hath these Waters, will without all
doubt compleat the Art.

Becker

Dr. Christian Augustus Becker describes his rediscovery of
what he calls the "Acetone of the Wise" in his work *Das
Acetone,* written in the German language, and published in
1862.

Becker was a physician and a student of the works of
Paracelsus. From his research into alchemy, he concluded that
there must be a solvent (menstruum) which was known to the
alchemists, able to work on minerals and metals, that proves
itself to be harmless to the human constitution.

155

> I had hoped for additional information in Weidenfeld's writings "De Secretis Adeptorum", but the main theme, the "Spiritus Vini Lullianus", remained a mystery.

After twenty years of study and work he found the key he was searching for.

> Thus the secret of the Spiritus Vini Philosophici was discovered and all the products from the distillation were correct. Now the Aqua Ardens with the quintessence became a simple chemical fact, and the only thing surprising that was left was how the old chemists had been able to work with it for centuries without this becoming known.

He had established for himself that what is known as the alchemical Mercury in the mineral world is acetone derived in a special way and is different than commercial acetone.

Based on his medical practice, Becker was of the opinion that the pure acetone, as provided by the chemical industry, is of little medical value because it is devoid of the subtle essential oils.

For medical application purposes, he advises that it be prepared with the same method used by the alchemists so that the oils remain intact.

Becker describes the distillate as mainly acetone containing two oily substances: one, a distillate at 90 deg; the other one, at 120 deg. These two oils form the core of the medication; therefore he called the substance "Acetonium Oleosum" or "Acetonol".

> The wine spirit is chemically always the same, but technically and physiologically it is different depending on its preparation from grain, rice, potatoes, wine, etc.; the same holds true for the acetone depending on the various bases of the acetate salts.

Commercial acetone production

The feverish interest in acetate distillation seen in the 1600s began to wane in the 1700s. As more physicians relied on apothecaries to prepare their medicines for them, experimentation with alchemical medicines also saw a decline, while the rise of the new science of chemistry ushered in processes for mass production of chemical materials. While acetone derived from the distillation of acetates became an important raw material of the emerging chemical industry, interest in it shifted to determining its chemical composition and structure, which was accomplished by the mid-1800s.

By 1888 the first large scale production of acetone in the United States was started in Albany, New York, by which time its alchemical origins had been almost forgotten.

Today, most acetone is produced as a byproduct from the manufacturing of phenol and to a lesser extent from the catalytic dehydrogenation of isopropyl alcohol.

Needless to say, this commercial product is not the Philosophical Wine Spirit of the adepts.

MATERIALS PREPARATION

Mineral and metal sources

The first task on this path is to select a mineral that contains the metal of interest. Mineral ores are basically salts consisting of an acidic metal species connected to an alkaline non-metal species. The following table lists some of the most common types of minerals from which the metals are obtained.

Metal	Oxide	Sulfide	Carbonate	Sulfate
Lead	Litharge	Galena	Cerussite	Anglesite
Tin	Cassiterite			
Iron	Hematite	Pyrite	Siderite	Green Vitriol
Copper	Cuprite	Chalcocite	Malachite	Blue Vitriol
Mercury		Cinnabar		...
Silver	Native Metal or in combination with copper, gold, lead, arsenic, or antimony			
Antimony	Senarmontite	Stibnite		
Gold	Native Metal only, or in rare combinations with silver and tellurium. Gold is found in minute quantities in almost all rocks and seawater			

Most Common Ores of the Metals

Once we have selected the mineral or metal for this work, it must be converted to a suitable form which will facilitate preparation of the acetate.

Preparing ores

In the alchemical view, mineral and metal ores collected fresh from the Earth are in their natural living form and preferred to any commercial chemical substitute.

Today, people don't usually consider rocks and minerals to be alive; but to the alchemist the mineral realm is just as full of life as the other two kingdoms of Nature. Minerals are alive; they grow, evolve, produce seed, and die just like plants and animals, but they do this at an imperceptibly slow rate.

Take the mineral ore you have selected to work on and separate by hand as much of the "high grade" material as possible from any surrounding rock.

Fold the collected material into a piece of cloth or canvas to avoid flying debris as you strike it with a hammer. Clean out any impurities that may have been freed, then continue crushing and grinding to a fine powder.

Very hard stones are often heated quite hot, then thrown into cold water a few times to help in reducing them to powder.

Most of the minerals have associated with them certain impurities which we want to get rid of as much as possible. Foremost of these are the more volatile metals which may include traces of arsenic, mercury, cadmium, selenium, zinc, and free sulfur which are slowly driven off by roasting the powdered ore outside for a day or two at about 90 deg C. Stir the material around once in awhile during the roast.

After this initial roasting, you can raise the temperature slowly to about 250 deg C for another day or two in order to remove still more traces of arsenic. It goes without saying that these are not things you want to be breathing, so have good ventilation.

Most of the common ores are easily converted to the oxide form by continued heating. The powdered ore is slowly calcined at higher temperatures, ground and calcined again.

The sulfide and sulfate minerals are often treated this way by slowly calcining the sulfur out, which is replaced by oxygen to form the oxide of the metal.

Carbonate minerals are easily calcined to the oxide with the release of carbon dioxide; they are also soluble in acidic solutions and form the salt of that acid. As an example, solution of lead carbonate in nitric acid forms lead nitrate; while solution in acetic acid will form lead acetate.

Many low grade minerals can be purified by solution in one of the mineral acids like nitric or hydrochloric acid, which separates the metal from the silica or alumina matrix.

The resulting acid solution is neutralized with sodium or potassium carbonate in solution. This will form the carbonate of the metal which is insoluble in water so it precipitates to the bottom and is easily collected. Some operators avoid this if possible, saying the corrosiveness of the acid will kill the alchemical life of the subject, but the mineral world is born of corrosive vapor.

If the acids are "philosophically" prepared according to the dry methods for distilling aqua fortis or strong waters, the effects are not so harsh. Hydrochloric acid, also known as

Muriatic acid, is said to be so filled with the life force, it can actually revive dead metals. Its preparation from sea salt is illustrated later in the "Book of Gold".

Once we have isolated the metal as a carbonate or oxide, we can macerate or reflux the material with a strong vinegar to form a solution of the metal acetate. Filter the solution and gently evaporate about 2/3 of it, then let cool undisturbed for a day or two covered with a thin cloth to keep dust out. During this time crystals of the metal acetate should form. Decant the liquid and return it to the metal oxide or carbonate along with fresh vinegar in order to extract more acetate. The crystals which formed are collected and saved in a sealed jar for later recrystallization. Repeat the process of extraction, concentration and collection of acetate crystals a second and third time.

All of the metal acetates are water soluble and with only a few exceptions can be recrystallized using rainwater. Iron acetate converts to a basic form which is a hydroxy acetate but still suitable for our needs, and mercury acetate tends to decompose into an oxide if left standing too long. Both can be recrystallized from alcohol if desired.

Dissolve the crude acetate into rainwater and then filter the solution. Gently evaporate the filtered solution to obtain purified crystals. This process can be repeated several times in order to produce very clean, shining crystals of the metal acetate. Allow the crystals to dry thoroughly, then store them in a tightly sealed container until you are ready for the distillation.

In the process of making the metallic acetates, we are altering the metal's body; moving its essence from the inorganic realm into the organic realm, preparing the metal to receive the vegetable nature and elevate the metal's essence.

It is the death and corruption of one body and rebirth into a new body which is improved and whose evolution is accelerated.

In the mineral realm, metals are generally associated with oxygen as oxides or with sulfur as the sulfides and sulfates.

Sulfur plays an especially important role in the mineral world similar to that played by oxygen in the plant and animal worlds.

Carbon forms the backbone of the organic realms and as we process our mineral into a carbonate or ultimately into an acetate, the metal obtains experience with carbon. This sets the stage for a whole new type of expression for the metal.

Metallic Lead and Carbon share an interesting relationship as they are both members of the IVa group of elements on the periodic table. Most of the ancient texts related to the acetate work place an emphasis (openly or in veiled terms) on the use of lead as the "matter" to be taken in hand.

Although silicon, germanium and tin are also members of the IVa group of elements, they are very difficult to work with in forming stable acetates. Only lead shares similar chemical properties with carbon which make it the ideal metal to assume organic life transferred from the vinegar.

Once we have prepared and purified the metal acetate, it should be sealed in a glass vessel, then placed into a warm incubator for a period of digestion and maturation of the vegetable life within the metal. The temperature should be around 40 deg C and the time should be at least a month, but longer is better. During this time you may observe the matter "sweating" within the glass and it may swell or continue growing new crystalline forms. These are all good signs that the vegetable life is "taking" within the metal.

The alkali metals, sodium and potassium, and the alkaline earths, calcium and magnesium, are good choices for initial experiments with the acetate method. These form relatively safe, non-toxic acetates representing important metals of the animal and vegetable worlds.

The sources for these metals are cheap and easy to come by. They are also easy to prepare and purify so you will be able to develop your skills in manipulating the materials and equipment necessary.

Natural sodium carbonate occurs as the mineral Natron from dried lakebeds. Natron was the primary drying agent used by the ancient Egyptians for preparing mummies.

The Salt of Tartar obtained from plant ashes or wine lees is potassium carbonate. This salt has many uses in practical alchemy and no lab should be without a good store of it.

Calcium carbonate is from limestone, and we can also use eggshells or seashells.

A combination of calcium and magnesium carbonates occurs in Dolomite Lime, which is sold at most gardening supply stores.

In practice you will find that these alkaline metals produce much more of the oils or alchemical Sulfur, along with their volatile spirit, than the more acidic metals like iron, zinc or lead. This also makes them good choices for beginning experiments as you learn to rectify the distillation products, a subject we will come back to later.

Wine Vinegar and Radical Vinegar

In order to prepare the metal acetates, we need a source of vinegar and this should be from a source we can trust as being "philosophical". Our main concern here is to preserve the vegetable life within the vinegar.

Vinegar was known to the Egyptians, and Caesar's armies drank it. Hippocrates prescribed the drinking of vinegar for his patients in ancient Greece. In all the places that we have seen the production of wine or beer in the ancient world, we also find the production of vinegar.

Vinegar is nothing more than an alcoholic beverage which has gone sour. In fact, that is exactly what the roots of the word mean, coming from the French *vin*, meaning wine, and *aigre*, meaning sour. Vinegar is made by the oxidation of alcohol, either directly or through the medium of a ferment, or by the distillation of wood; the latter is known as pyroligneous acid and is not suitable for our work here. Any substance capable of fermentation or any containing alcohol is suitable for making vinegar. When alcoholic beverages sour, it is the action of certain bacteria, known as acetobacter, on alcohol, turning it to acetic acid and water. It is the alcohol gone acid which gives us the taste that we associate with vinegar. It is the other elements, specific to the actual source of the original alcohol, which give the vinegar its individual character and body.

The acetobacter reaction, unlike that of yeast on sugar to make alcohol, is an aerobic reaction. It requires the presence of oxygen. The more oxygen, the better.

In today's market, there are many adulterations of Vinegar. Sulfuric, nitric and hydrochloric acids are used to give a false strength; burned sugar and acetic ether to give color and flavor; various preservative chemicals to extend shelf life.

If possible it is always preferred to make your own vinegar so you know exactly what has gone into it.

Vinegar made from red wine is held to be the best for alchemical works. To make wine vinegar you want good quality wine that's not too strong, 10-11% alcohol at the most, because too much alcohol inhibits the activity of the bacteria that transform the wine. If the wine is too weak on the other hand, less than 5%, the vinegar won't keep well.

The simplest way to proceed is to leave an open, 3/4 filled bottle of wine in a warm place for a couple of weeks to a month. The optimum temperature for this conversion is about 29 deg C.

For a steady supply of vinegar, take a wide-mouthed glass jug whose capacity is at least a gallon and pour a quart of wine and a cup of vinegar into it. Keep the container covered most of the time, but open it for a half hour every day. In a couple of weeks the "mother", a viscous starter, will have settled to the bottom of the jug, while the vinegar above it will be ready for use.

Add more wine as you remove vinegar to keep the level in the jug constant. Replace the vinegar removed with more wine, pouring it into the container with a length of hose so as to leave the surface molds undisturbed. The reason for this is that a scum will form on the surface of the liquid as it is converted to vinegar. This is a very active layer of acetobacter and forms on the surface, where there is the most oxygen (from contact with the air). While succeeding batches of vinegar will proceed even if this layer is broken up, they will get off to a much better start if this layer is left undisturbed.

Once we have a good quantity of vinegar, our next step is to concentrate its power to dissolve the minerals. Although it will work as produced above, it will require much more volume and time to do its work effectively.

The vinegar is concentrated first by freezing and later by distillation. For this you can half fill a plastic bottle with vinegar and freeze it. Take it out of the freezer and overturn it into a glass jar. After 30 minutes or so, the concentrated vinegar will have thawed and dripped into the jar, leaving a plug of ice in the plastic bottle. Discard the ice, then refreeze the vinegar and repeat this thawing/freezing cycle yet a third time.

The concentrated vinegar can be used at this point without further preparation, and you can be confident that the delicate plant life has been preserved because there has been no exposure to high temperatures. Some operators prefer to gently distill the concentrated vinegar but discard the first 1/4 to 1/3 as it is mostly water. This will produce a very strong and clear vinegar, but it is best to perform this distillation under vacuum so the lowest temperature can be used. Distill to near dryness and collect the distillate. This is the Fixed Vegetable Spirit.

There is another way to make a highly concentrated vinegar and though it requires much more work, it will already be loaded with mineral fire united with the vegetable life and thus readily accepted by any of the metals.

This is called the "Radical Vinegar" and it provides a solvent whereby most of the mineral and metallic realm can be opened. The starting materials are easy to find, copper wire and red wine vinegar.

The copper wire is balled up loosely and heated to redness several times in order to oxidize it to a black color.

Place the brittle mass into a glass vessel and cover with wine vinegar which has been concentrated by freezing. Seal the container and let it digest and shake it at least daily. After some time, the liquid will become deep emerald green. Decant the liquid into another vessel and save it aside. Repeat the process of heating the wire and extracting with fresh vinegar several times. Collect all of the liquid extracts together and filter it into a porcelain dish. Evaporate the liquid gently and collect the deep green crystals of copper acetate that form. These can be recrystallized from rainwater several times to enhance their purity.

The dry crystals are crushed, then placed into a strong distillation apparatus.

With a cooled receiver in place, the crystals are heated first slowly, then gradually up to about 400 to 500 deg C. Here again the fumes can be quite overpowering, so be sure to have adequate ventilation. The liquid distillate is a very concentrated acetic acid and may have a slight blue-green tint. This is the Radical Vinegar.

With either of these strong vinegar solutions, we can proceed to make the metal acetates as detailed above.

Apparatus

This method requires the use of what is called "dry distillation", where a solid is placed into the stillpot instead of a liquid and distilled at fairly high temperatures.

The first time I saw the dry distillation of a metal acetate used a simple test tube, heated by a propane torch, with a bent piece of tubing into a second test tube on ice. No doubt we lost a great portion of the more volatile Mercury but were successful in obtaining a volatile distillate from which an oil and flammable spirit could be collected. Anything from here is an improvement as far as apparatus goes, but it demonstrates how readily available the method is for anyone to experiment with.

Let's look at some apparatus designs from the ground, up as it were, beginning with the support system. You need to be able to support the apparatus firmly but not too rigidly, as there will be some expansion and contraction of the system.

A very serviceable way to accomplish this is to construct a lattice consisting of 1/4 to 3/8 inch iron pipe cut into three- or four-foot lengths and lashed together with strong iron wire. The end posts are attached to the benchtop using pipe floor flanges, which are available from a plumbing supply store.

Equipment Lattice

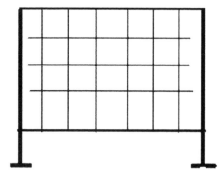

After the lattice is attached to the benchtop, it is a good idea to attach a safety wire from the top rail to the ceiling in order to lend additional support for top-heavy loads.

Heating

Heating for the flask containing the metal acetate is provided by gas or electric means. The temperature must be able to hit the range of 400 to 700 deg C. There are high temperature mantles, which fit snugly around flasks, but they tend to be quite expensive. Hot plates or small portable stovetop ranges are also available which are very serviceable. The flask can be buried in a sandbath to provide an even heating all around the vessel and lend support to the glass which may tend to deform at the high temperatures.

Gas heating from bunsen burners, gas ranges, or propane torches are also an option which allows good control of the temperature; however, be aware that the distillate and gasses evolved are flammable, and some are toxic, so provide good ventilation and have a fire extinguisher ready if things get out of control. Never leave the distillation unattended.

Initially, the temperature will be relatively low (about 100 deg C) at the distilling head as the Phlegm comes over. Once this passes, the temperature is slowly raised until the white vapor rises and fills the apparatus. When the white vapor begins to diminish, increase the heat even more until the red oil distills over. Once distillation seems to be at an end, gradually turn the heat down, then finally off, and let cool.

While the use of thermometers is helpful, they are not essential to success. Mercury thermometers are only useful to about 350 deg C, so if you want to monitor the sandbath temperature, it is better to use a thermocouple and meter. Type K thermocouples from the pottery supply shop are suitable for this.

Glassware simple and complex

As we mentioned earlier, the glassware involved in the acetate distillation can be surprisingly simple or more complex in order to minimize losses of the volatile spirits.

In the old days, many artists would attach a very large receiver to the distillation flask and seal the joint with a "lute" composed of cloth strips smeared with a mixture of clay, iron filings, and egg whites. The receiver was sufficiently large such that the acetate being distilled could entirely vaporize and there would still be enough room to retain it without building up tremendous pressures that could burst the vessels.

Today, the most common way of performing the distillation is with an apparatus consisting of three parts (see diagram below). First is the "still pot" or flask containing the acetate to be distilled (A). This is the part that takes the heat and suffers the most damage. This flask is attached through a condenser to a "first receiver" (B) which will collect the bulk of the distillate. This flask should have the option of being easily exchanged with a fresh flask or be provided with a drainage stopcock. The first receiver is, in its turn, connected to a third flask, called the

167

"cold trap" (C). The cold trap provides an especially low temperature cooling environment to capture the most volatile portions of the spirit.

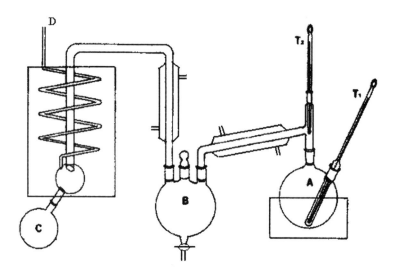

This setup presents an ideal, but there are options such as replacing flask A with a simple erlenmeyer flask and attaching it with a dense cork or silicone stopper to the rest of the distillation train. This way you won't have to worry about burning up a nice ground glass flask after several runs. The thermometers are optional.

Flask C can be immersed in a container of ice water or one of the cooling mixtures we will talk about next. The venting tube D can be attached to a bubbler train, which will allow you to capture the very volatile spirit into a suitable medium.

The main objective is to provide enough cooling surface to slow down and condense the vapors that will be streaming through the apparatus.

The illustration above is from *The Art of Distillation* by John French, published in 1651. The apparatus depicts the same essential parts of the acetate distillation train.

Cooling and use of Bubblers

The purpose of the cold trap is to provide a very cold environment, which will cause the vapor to condense as much as possible. At the very least it should be a slurry of ice water that surrounds flask C. There are a few simple methods we can use in order to make this cold trap even colder. A mixture of one part common salt and three parts crushed ice will give a temperature in the range of –5 to –18 deg C. If we use a mixture of five parts calcium chloride with four parts crushed ice, we can get the temperature down to –40 to –50 deg C, which is sure to condense most of the vapor. Calcium chloride is sold in many hardware stores for use in room dehumidifiers.

Many operators prefer to use a dry ice and acetone bath, which can reduce the temperature to nearly –80 deg C. The advantage to this is that the materials are inexpensive and you don't have the mess of salt mixtures. The disadvantage is that the acetone vapors are flammable, so gas heating is not advisable; use electric heating only.

The bath is prepared by partially filling the cold trap vessel surrounding flask C with commercial acetone from the hardware store. Small pieces of dry ice, which is solid carbon dioxide, are very slowly added to the acetone. Use gloves, tongs or pliers to handle the dry ice, as it can cause severe frostbite to the bare hands. At first, the acetone will bubble violently, so be careful. As the liquid cools, the bubbling will become quiet and larger pieces of dry ice can be added to maintain the low temperature. After completing the distillation, save the acetone as it can be used for this purpose again.

Even with these very icy temperatures, the spirits rising from the minerals can be subtle and hard to capture. Another useful approach to catching these fugitive spirits is the use of what is called a bubbler train. For this, the vent line from the distillation train is connected to a series of bottles set up as shown in the diagram below.

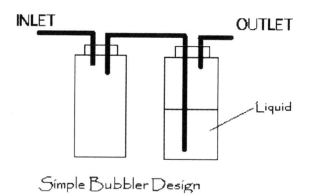

Simple Bubbler Design

The first bottle is left empty and acts as a safety container in case there is a sudden drop in pressure. The second bottle is partially filled with alcohol or acetone. As the distillation proceeds, the subtle vapors which are not condensed bubble through the liquid and become dissolved into it. We can even add a third and fourth bottle to the train to be sure we capture all of the spirits. If the pressure suddenly drops inside the apparatus, the liquid would be sucked into the system if it were not for the empty first bottle acting as a catch vessel.

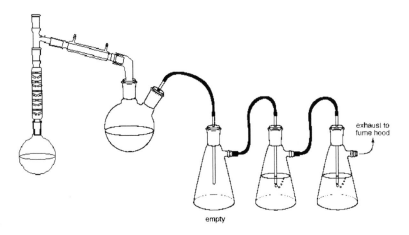

empty

exhaust to fume hood

THE PROCESS

Before describing the process of distilling the acetate and subsequent rectification of the products, here is another excerpt on the process, which is held in high esteem for its clarity, written by the French alchemist Joseph Du Chesne, who was also known as Qercetanus. This piece contains valuable keys to the entire acetate work.

> Treatise on Metallic Medicine by Joseph Du Chesne, being "A Collection of the Most Precious and Rare Secrets, Taken from the Manuscripts of the late Monsieur Joseph Du Chesne, Sieur de la Violette, Officer and Physician in Ordinary to the King", Paris, 1641

> Preparation of Saturn which is Efficient against the Lepra of Human and Metallic Bodies, and of which an Oily Solvent can be made.

> Distill a large quantity of good vinegar, till you have a cask full of it, because it is the basis and the foundation of this Work. To strengthen it

171

more, distill it several times over the feces, then mix everything you have distilled with as much other non-dephlegmatized vinegar, and let them go over together, so that the distillate will become all the more efficacious. The dregs that remain at the bottom are put in a retort over a good fire, by means of which one can extract an excellent oil from them, which can burn of its own and dissolve all kinds of minerals.

After preparing this solvent, take 80 lbs of powdered litharge—and NOT white lead or minium of lead calx (oxide), as several artists do, especially Isaac Hollandus. Take, I say, this litharge and put it in several big and very strong flasks. Pour on it as much of your vinegar that it will overfloat by 6 fingers' breadth, and then put it on an ash-fire. Extract the salt of Saturn by a slow digestion, and on the feces that are left after the extraction of the salt and the crystals, pour once more the same amount of menstruum as indicated above. Continue doing this till all your litharge has turned into crystals which are, properly speaking, what the philosophers call the Chaos or the metallic materia prima.

On this crystalline substance, again put for the last time fresh distilled vinegar. Dissolve it over a slow fire and filter it, so as to obtain a perfectly pure and flawless menstruum which, after passing through the steam-bath, will leave a substance that melts like wax at the bottom of the alembic. It hardens in the cold as it melts in the heat. Thereafter, divide this melting substance among several alembics and little by little pour fresh menstruum upon it, as if to feed and water it only. Do this by first pouring on only two drops, then three, then five, then seven, increasing the amount in this way till the materia does not absorb any more. You will recognize this when you see the solvent coming out as acid as it was at the beginning. Therefore, whenever you distill your imbibed materia, take care that you continue till the phlegma is as acid as before, because this is how the child refuses the nurse's milk when its stomach is full. When the materia has been prepared in this way and converted into an excellent and precious gum, digest it in the steam bath for 30 or 40 days, till it becomes black and has a bad smell like that of liquid pitch. It is from this liquid and black pitch that you must extract, by the same bath, an excellent phlegma which can serve as a proper menstruum for extracting a precious salt from calcined earth, as we will write later. Owing to the continual distillation that you will make of the said pitch on sand, and by finally giving a strong fire above and below through the usual degrees up to a very violent fire, you will extract a red and quite thick oil which, together with the preceding distillations, will constitute as strong and violent a water as that extracted from wine, and will have the same great power. The philosophers call it water of life (brandy) of Saturn. Its substance is so pure and subtle that it must be kept in a well-closed vessel lest it evaporate.

To complete the perfection of this solvent, this water of life of Saturn must be put in a gentle bath, in a long-necked alembic, where the purest spirit of this water will rise imperceptibly till you see the appearance of some lines and filaments through the glass of the head. It is an infallible sign that all of the spirit has risen, and you must therefore stop this distillation and extract this first precious spirit. Preserve it carefully in a cold place and in a well-sealed container. After this spirit, a milky phlegma will appear in a stronger bath. It will be much better for washing your calcined materia than the first of which we spoke above. Finally, by a stronger degree of fire and after changing the receptacle, you will still separate an ardent spirit which will first come out white and watery, then red and oily, but it will be heavy and lie at the bottom of the receptacle. However, if you wish, you can make it go over with a stronger fire.

In regard to the earth or the feces that are left at the bottom of the retorts as a black powder, they can also be dissolved with some fresh distilled vinegar and thus turn into new lapilli of a sticky and gummy consistency, and finally, by means of the above-mentioned digestions and distillations, into wonderfully active and burning spirits. There are some who divide this earth into two, but although Isaac (Hollandus) himself adopts this division, I am nevertheless of the opinion that the best and shortest method is to calcine all the earth together and to reverberate it by a gentle flame till it becomes yellow like ochre. When this earth has become yellow due to the cohobation of the phlegmas, the salt can again be separated from it, according to the ordinary rules and operations of the Art.

Having achieved the extraction of this rare and precious salt, take the first salt which you have little by little extracted and which you have preserved. Pour it on 1 oz of the last salt, repeating this imbibition till 1 oz of this salt weighs 3 or 4 oz and has retained the weight of the sal ammoniac of this spirit, till finally the volatile exceeds the fixed. If you work this process exactly, you will find an excellent earth at the bottom. Sublimate it in a very clear and well sealed glass vessel, and you will have the pleasure of seeing in it the sublimation of a Philosophical Mercury in the form of a fine talc, which you must keep as a most valuable substance.

To crown this work, take 1 part of this Mercury and add it to 4 parts of the above-mentioned spirit or to as much ardent spirit to make of them a solvent for the Sun and the Moon, such as the philosophers imagined were capable of turning them into spirit without destroying their bodies or losing their characteristics. Therefore, wonderful works can be made with this truly philosophical solvent, both for the health of human and of metallic bodies.

Dry distillation

This is where the real fun begins. The metal or mineral ore we have selected to work on has been converted into its acetate form and purified by recrystallization. This is now our matter which is to be distilled in this crystalline state. The crystals should be very gently dried and crushed like coarse sand, then placed into the distilling flask. Do not fill the flask more than about one third full, as many of the acetates will expand or foam during the heating process. If foaming becomes a problem, the sages recommend saturating granulated, porous brick or pottery shards with the acetate solution and letting them dry. This saturated material is then used for the distillation and helps to reduce foaming through a more gentle release of the vapors. This is especially recommended for potassium acetate, which tends to foam up a great deal.

Attach the flask to the distillation train and start the coolant flow through the condensers. If you are going to use an additional cold bath for the receiver, start cooling it as well. If you are using a bubbler system, make sure it is charged with alcohol or acetone.

When all is ready, start heating the distillation flask slowly to about 100 deg C. As the temperature is rising, moisture will begin to collect and come over into the recipient. This is mostly water that was associated within the crystalline acetate, the so-called "water of crystallization". Collect this water and save it aside; this is called the "Phlegm", and is used to extract the salts later. As the heat increases, the phlegm will stop distilling over; it is convenient to change receivers at this point or drain the first receiver of the phlegm if you have a stopcock attached.

Depending on the particular acetate being distilled, the matter in the flask may begin to melt, or if already liquified, it may suddenly solidify and begin to swell. With a fresh receiver in place, continue heating. Soon you will see a thick, heavy, white vapor coming over and a golden liquid form in the receiver, which must be kept very cold or this spirit will escape.

The temperature at the distilling head may level off even though the flask temperature is still rising. The final distillation temperature may be in the range of 400 to 700 deg C, and by this time you will notice drops of blood red oil distilling over; the apparatus being filled with dense white vapor so as to appear made of chalk. After some time the vapor will diminish and distillation will stop. Detach the bubbler train if you used one and let the apparatus cool. Finally remove and save the distillate in a tightly sealed container. This liquid may be golden or deep red like wine and is the Menstruum Foetens of Ripley or Secret Wine Spirit of Weidenfeld.

The black residue remaining in the flask is the Black Dragon. During the distillation, some of the salts may sublimate and deposit themselves in the upper portions of the distillation flask. Save these salts aside separately, as they are a valuable treasure for later use.

Rectification

After the distillation is complete, we will have several products collected and saved separately. These include the phlegm collected first, the distillate in the first receiver which is the menstruum foetens, and the cold trap distillate which is the most subtle portion of the aqua ardens. If we also used a bubbler train, that liquid contains some of the aqua ardens as well. And finally there is the residue left in the distillation flask which represents the Salt of the metal.

The phlegm can be purified by gentle distillation several times. It possesses a solvent power and medicinal virtue in its own right, due to its intimate association with the metal it was derived from.

The menstruum foetens contains the red and white philosophical wine spirits. This liquid has power to dissolve many of the metals, even gold if it is prepared in a finely divided state. By a very gentle distillation, the white wine spirit or aqua ardens will separate and come over at about 56 deg C. It can be matured by gentle digestion and redistillation several times.

This liquid represents the Philosophical Mercury of the metal. During these digestions and distillations, a clear to golden yellow oil may separate from the bulk of the liquid; save it aside separately, it is the Quintessence of the White Wine Spirit.

The red wine spirit is saved aside separately at this point, or it can be reduced into a thick blood red oil by distillation at a higher temperature. In this distillation, a clear acidic liquid generally distills over first, followed by a yellow oily liquid. The red oil remains in the distillation flask and represents the Philosophical Fixed Sulfur of the metal. The oils derived from the acetate distillation are not simple compounds but instead are very complex mixtures of many constituents, much like the essential oils derived from plants.

Another way to separate the red oil from the red wine spirit is by extraction with ether. First the aqua ardens must be separated by distillation as described above.

A small amount of ether is then added to the red wine and mixed by shaking well. Be sure to relieve the pressure once in a while during this extraction and also be sure there are no flames present, as ether is explosively flammable. After shaking, let the liquid settle and the ether will form a layer at the top which is darkly colored red. Draw off this ether layer and place it into a container. Let the ether evaporate and the red oil will remain in a purified state without having to subject it to the high temperatures required for its distillation.

The solid residue remaining in the distillation flask, our black dragon, is generally in the form of an oxide or carbonate. It can be calcined and turned into an acetate again, followed by another distillation to obtain more of the menstruum foetens, or it can be extracted with the phlegm and crystallized to obtain the Salt.

In the case of lead acetate, the residue is a finely powdered lead metal. Be sure the flask is cool before it is removed. This fine lead powder is so anxious to oxidize that upon contact with the air it will spontaneously calcine itself, like the phoenix, and turn into a beautiful yellow to red lead oxide. If it is poured out of the flask while still hot, it may burst into flame and even shatter the flask.

Uses

Now that we have collected the various products of the distillation, we must decide how to use them. As in the vegetable work, these products represent the three essentials of the particular metal used. Some operators use only the Mercury and Sulfur. After purification these are combined to produce "Philosophical Gold" which is then diluted with alcohol to form a tincture with powerful medicinal virtues related to the planet of their origin.

From an ayurvedic perspective, the Philosophical Mercuries represent the vata part with drying, cold, mobile, subtle qualities. The metallic oils, the Sulfur, carry the qualities of pitta, being oily, hot, sharp, and mobile. The Salt carries the qualities of kapha, which are cold, heavy, static, gross, and dense. The different proportions and components found in the various minerals each express a particular blend of the gunas.

Weidenfed presents comparisons from different alchemical writings for the preparation of menstruums by "sharpening" the distillates with various materials.

He describes twenty-four classes of menstruums useful for extracting the essences from vegetable, animal, or mineral subjects. The most simple of these are classed as vegetable menstruums and they are prepared by circulating the aqua ardens with various plants, especially hot, spicy plants.

… the Matter of the Menstruum of Vegetable Mercury or Soul of Metals, is not Common, but Philosophical Wine; nor that the Spirit of this Wine is the Common, but Philosophical Aqua ardens.

A Menstruum of this kind is the unctuous Spirit of Philosophical Wine acuated, that is, tempered with the common Unctuosity of Vegetable Oyls. Mix, digest, and distil any common distilled Oyl with the Spirit of Philosophical Wine, and you will obtain a Menstruum of the Second Kind much sooner.

Each of the menstuums he describes also possesses a unique medicinal virtue in its own right as well as the power to reanimate materials, even smelted metals, imbibed with them.

Aqua vitae is the Soul and Life of Bodies, by which our Stone is vivified. So also Ripley in Libro Mercurii, saith, The Sperm of Metals is also called Metallick Aqua Vitae, because it administreth life and health to Metals, being sick, dead, etc.

For Menstruums are the Souls of Metals, by which the Metals, otherwise dead, are animated and revivified.

Other substances used to sharpen or acuate the Philosophical Wines include: volatile salts like sal ammoniac; fixed salts like salt of tartar and sea salt; volatile acids like hydrochloric acid and other "salt essences"; various forms of aqua fortis; and fixed mineral acids like sulfuric acid or oil of vitriol. Various combinations of these materials are also used to create compound menstruums.

The general process of sharpening involves digestion of the substance with the wine spirit, followed by distillation. The distillate is returned to the residue, or fresh material is used, and the process of digestion and distillation is repeated several times.

The completed menstruums are used medicinally on their own, or in the further preparation of metals to be used medicinally and in experiments of transmutation.

Some of the medicinal claims for these materials, as found in the old texts, border on the miraculous.

In *Das Acetone*, Becker mentions the work of Count Onuphrio de Marsciano, who describes in his hermetic writing of 1774 how he cured himself of gout by external application of the distillate. The Count also recommends its use internally: "Take 20 drops in the morning before eating for 15 days in order to completely clean the blood since there is no blood cleansing like it in the whole world". He calls the substance spiritus simplex, and cites from Lully that "the quintessence heals all tiredness and sickness, removes all weakness, protects from all sicknesses and retains the youth, he clearly says: 'And I swear the truth that I have seen wonderful things done by this Simplici Spiritu Vini philosophici, and I have even healed gout completely with it.'"

As a medical doctor, Becker recorded his own observations on the medicinal uses of the rectified wine spirit which he calls "Spiritus Aceti Oleosus".

He summarizes a number of case histories wherein the preparation was successful in the treatment of gout, rheumatism, edema, colds, flu, infected wounds, gonorrhea, neuralgia, sciatica, and fever.

A typical prescription of his contained 3.5 cc spiritus aceti oleosus dissolved into 2oz. Distilled water and sweetened with ½ oz. of syrup. The dose being one tablespoon every 2 to 3 hours.

Through his practice he was convinced that this spirit has great utility in medicine and urged his fellow physicians to regain possession of this valuable remedy.

Once the separation and purification of the three essentials has been performed by distillation of the acetate, most of the old texts instruct us to imbibe the purified Salt with the Mercury and Sulfur as is done when making the Vegetable Stone.

The black residue is calcined, the salt extracted and crystallized. Then it is soaked with the etheric spirit. Through sublimation you obtain the "Terra Folliata Philosophorum", which has a shine stronger than Oriental pearls.

When the red oil is added to this Terra Folliata and combined with it through repeated cohobation and distillation, there results the true solvent of nature and the quintessence of magnificent power: this quintessence is the true and living, clearest source in which Vulcan washes Phobus (gold), and cleans of all impurities and creates the means to fortify the strength of life, improves everything weak, and renews the power of youth.

Das Acetone, Becker on lead acetate

As Ripley, quoted above, says:

... in the end you will have "The True Oil and Elixir, ...which being duly prepared, doth not only alter and change the filthy and corrupt humors of our bodies, but also can change and transmute Luna into Sol".

Analytical Data

See appendix IIA for a selection of analytical data collected on some of the products obtained from the acetate path.

CHAPTER SIXTEEN

The Book of Antimony

What is antimony? It's not a word you hear everyday and as a material, it's not generally well known either. Today we find antimony used in batteries, matches, and electronics. It's used in metal alloys to provide a degree of hardness required for bearings. Indeed it can make metals glass hard and just as brittle as itself. As a metal it is pretty easily ground into powder, unlike the other metals we are familiar with. Did I mention it's toxic? Quite deadly really, like its close associate Arsenic. Antimony is a very odd material with a thousand uses in modern technology; it has an equally odd history filled with secrecy, mystery, miracles and scandal.

History

Archeological evidence points to man's knowledge of antimony as early as 4000 BCE, not only as a mineral but as a metal used to cast vessels and small ornamental items.

Ancient Egyptians used finely powdered stibnite, which they called Mestem, as eyeliner for both cosmetic and medicinal purposes as mentioned in the Ebers medical papyrus of about 1700 BCE. This fine powder was later called Khol, and you will still see eyeliner sold under this name. Khol was a term used for centuries to describe anything in a very finely powdered state. Later, Arabic influence changed this to Al-Khol, and still later it became known as Alcohol, meaning the finest powder. It was not until the middle ages that the term alcohol was used to designate the spirit of wine. It is interesting to note that the Arabic term Al-iksir, from which we get our word elixir, was also used to designate a finely powdered substance. The word is derived from the early Greek word xirion, meaning a dry powder or ash.

In the ayurvedic tradition, antimony is also prepared as a fine powder used to treat diseases of the eyes.

Antimony was also used to purify gold, a subject we will talk about later in "The Book of Gold".

Roger Bacon 1214-1294 CE

The alchemist Roger Bacon presents a clear description of the antimony work leading to medicines for man and metals.

Excerpts from "Tract on the Tincture and Oil of Antimony" by Roger Bacon.

> Stibium, as the Philosophers say, is composed from the noble mineral Sulphur, and they have praised it as the Black Lead of the Wise. The Arabs in their language, have called it Asinat vel Azinat, the alchemists retain the name Antimonium. It will however lead to the consideration of high Secrets, if we seek and recognize the nature in which the Sun is exalted, as the Magi found that this mineral was attributed by God to the Constellation Aries, which is the first heavenly sign in which the Sun takes its exaltation or elevation to itself.

> Yes, many say, that when one prepares Stibium to a glass, then the evil volatile Sulphur will be gone, and the Oil, which may be prepared from the glass, would be a very fixed oil, and would then truly give an ingress and Medicine of imperfect metals to perfection. These words and opinions are perhaps good and right, but that it should be thus in fact and prove itself, this will not be. For I say to you truly, without any hidden speech; if you were to lose some of the above mentioned Sulphur by the preparation and the burning, as a small fire may easily damage it, so that you have lost the right penetrating spirit, which should make our whole Antimonii corpus into a perfect red oil, so that it also can ascend over the helm with a sweet smell and very beautiful colors and the whole body of this mineral with all its members, without loss of any weight, except for the foecum, shall be an oil and go over the helm.

The Practica Follows:

Take in the Name of God and the Holy Trinity, fine and well cleansed Antimonii ore, which looks nice, white, pure and internally full of yellow rivulets or veins. It may also be full of red and blue colors and veins, which will be the best. Pound and grind to a fine powder and dissolve in a water or Aqua Regis, which will be described below, finely so that the water may conquer it. And note that you should take it out quite soon after the solution so that the water will have no time to damage it, since it quickly dissolves the Antimonii Tincture. For in its nature our water is like the ostrich, which by its heat digests and consumes all iron; for given time, the water would consume it and burn it to naught, so that it would only remain as an idle yellow earth, and then it would be quite spoilt.

Therefore you must remember to take the Antimonium out as soon as possible after the Solution, and precipitate it and wash it after the custom of the alchemists, so that the matter with its perfect oil is not corroded and consumed by the water.

The Water; wherein we dissolve the Antimonium, is made thus:

Take Vitriol one and a half pound, Sal armoniac one pound, Arinat (Alum) one half pound, Sal niter one and a half pound, Sal gemmae one pound, Alumen crudum one half pound. These are the species that belong to and should be taken for the Water to dissolve the Antimonium.

Take these Species and mix them well among each other, and distill from this a water, at first rather slowly, for the Spiritus go with great force, more than in other strong waters. And beware of its spirits, for they are subtle and harmful in their penetration.

185

When you now have the dissolved Antimony, clean and well sweetened, and its sharp waters washed out, so that you do not notice any sharpness any more, then put into a clean vial and overpour it with a good distilled vinegar. Then put the vial in Fimum Equinum, or Balneum Mariae, to putrefy forty days and nights, and it will dissolve and be extracted red as blood. Then take it out and examine how much remains to be dissolved, and decant the clear and pure, which will have a red colour, very cautiously into a glass flask. Then pour fresh vinegar onto it, and put it into Digestion as before, so that that which may have remained with the faecibus, it should thus have ample time to become dissolved. Then the faeces may be discarded, for they are no longer useful, except for being scattered over the earth and thrown away. Afterwards pour all the solutions together into a glass retort, put into Balneum Mariae, and distill the sharp vinegar rather a fresh one, since the former would be too weak, and the matter will very quickly become dissolved by the vinegar. Distill it off again, so that the matter remains quite dry. Then take common distilled water and wash away all sharpness, which has remained with the matter from the vinegar, and then dry the matter in the sun, or otherwise by a gentle fire, so that it becomes well dried. It will then be fair to behold, and have a bright red color. The Philosophers, when they have thus prepared our Antimonium in secret, have remarked how its outermost nature and power has collapsed into its interior, and its interior thrown out and has now become an oil that lies hidden in its innermost and depth, well prepared and ready. And henceforth it cannot, unto the last judgment, be brought back to its first essence. And this is true, for it has become so subtle and volatile, that as soon as it senses the power of fire, it flies away as a smoke with all its parts because of its volatility.

When you then have the Antimonium well rubified according to the above given teaching, then you shall take a well rectified Spiritum vini, and pour it over the red powder of Antimony, put it in a gentle Balneum Mariae to dissolve for four days and nights, so that everything becomes well dissolved. If however something should remain behind, you overpour the same with fresh Spiritu vini, and put it into the Balneum Mariae again, as said before, and everything should become well dissolved. And in case there are some more faeces there, but there should be very little, do them away, for they are not useful for anything. The Solutions put into a glass retort, lute on a helm and connect it to a receiver, also well luted, to receive the Spiritus. Put it into Balneum Mariae. Thereafter you begin, in the Name of God, to distill very leisurely at a gentle heat, until all the Spiritus Vini has come over. You then pour the same Spiritum that you have drawn off, back onto the dry matter, and distill it over again as before. And this pouring on and distilling off again, you continue so often until you see the Spiritum vini ascend and go over the helm in all kinds of colours. Then it is time to follow up with a strong fire, and a noble blood red Oleum will ascend, go through the tube of the helm and drip into the recipient. Truly, this is the most secret way of the Wise to distill the very highly praised oil of Antimonii, and it is a noble, powerful, fragrant oil of great virtue, as you will hear below in the following.

Take the mixture of oil and wine spirit put it into a retort, put on a helm, connect a receiver and place it all together into the Balneum Mariae. Then distill all the Spiritum vini from the oil, at a very gentle heat, until you are certain that no more Spiritus vini is to be found within this very precious oil. Then remove the fire from the Balneo, though it was very small, so that it may cool all the sooner. Now remove the recipient containing the Spiritu vini, and keep it in a safe place, for it is full of Spiritus which it has extracted from the oil and retained.

It also contains admirable virtues, as you will hear hereafter. But in the Balneo you will find the blessed bloodred Oleum Antimonii in the retort, which should be taken out very carefully.

Thus you now have two separate things: Both the Spirit of Wine full of force and wonder in the arts of the human body: And then the blessed red, noble, heavenly Oleum Antimonii, to translate all diseases of the imperfect metals to the Perfection of gold. And the power of the Spiritual Wine reaches very far and to great heights. For when it is rightly used according to the Art of Medicine: I tell you, you have a heavenly medicine to prevent and to cure all kinds of diseases and ailments of the human body.

Powder of Projection

Take, in the Name of God, very pure refined gold, as much as you want and think will suffice. Dissolve it in a rectified Wine, prepared the way one usually makes Aquam Vitae. And after the gold has become dissolved, let it digest for a month. Then put it into a Balneum, and distill off the spiritum vini very slowly and gently. Repeat this several times, as long and as often until you see that your gold remains behind in fundo as a sap. And such is the manner and opinion of several of the ancients on how to prepare the gold. But I will show and teach you a much shorter, better and more useful way. Viz. that you instead of such prepared gold take one part Mercurii Solis, the preparation of which I have already taught in another place by its proper process. Draw off its airy water so that it becomes a subtle dust and calx. Then take two parts of our blessed oil, and pour the oil very slowly, drop by drop onto the dust of the Mercurii Solis, until everything has become absorbed. Put it in a vial, well sealed, into a heat of the first degree of the oven of secrets, and let it remain there for ten days and nights. You will then see your powder and oil quite dry, such that it has become a single piece of dust of a blackish grey colour. After ten days give it the second degree of heat, and the gray and black colour will slowly change into a whiteness so that it becomes more or less white. And at the end of these ten days, the matter will take on a beautiful rose white. But this may be ignored. For this colour is only due to the Mercurio Solis, that has swallowed up our blessed oil, and now covers it with the innermost part of its body. But by the power of the fire, our oil will again subdue such Mercurium Solis, and throw it into its innermost. And the oil with its very bright red colour will rule over it and remain on the outside. Therefore it is time, when twenty years have passed, that you open the window of the third degree. The external white colour and force will then completely recede inwardly, and the internal red colour will, by the force of the fire, become external. Keep also this degree of fire for ten days, without increase or decrease. You will then see your powder, that was previously white, now become very red. But for the time being this redness may be ignored, for it is still unfixed and volatile; and at the end of these ten days, when the thirtieth day has passed, you should open the last window of the fourth degree of fire, Let it stay in this degree for another ten days, and this very bright red powder will begin to melt. Let it stay in flux for these ten days. And when you take it out you will find on the bottom a very bright red and transparent stone, ruby colored, melted into the shape of the vial. This stone may be used for Projection, as has been taught in the tract on Vitriol. Praise God in Eternity for this His high revelation, and thank Him in Eternity. Amen.

Multiplication of quantity

Take in the Name of God, your stone, and grind it to a subtle powder, and add as much Mercurii Solis as was taught before. Put these together into a round vial, seal with sigillo Hermetis, and put it into the former oven exactly as taught, except that the time has to be shorter and less now. For where you previously used ten days, you may now not use more than four days. In other respects the work is exactly the same as before.

Nicholas Flamel 1330-1417 CE

One of the most popular stories of a successful alchemist is that of Nicholas Flamel. He was a bookseller of modest means with an interest in the Hermetic Arts. He came into possession of a book filled with hieroglyphic figures, which he recognized as a valuable work on alchemy. After nearly thirty years of study he was able to understand the writings and figures in the book and undertake the practical work. Flamel was able to complete the Great Work of confecting the Philosopher's Stone and transmutation of metals. Though he continued to live modestly, he used his newfound wealth in charitable ways such as establishing several free hospitals for the poor, sizeable endowments to the church, and construction of housing for the homeless. His handwritten texts and full documentation of these acts are kept in the Bibliotheque Nationale in Paris.

His method for preparing the Philosopher's Stone is a Dry Way of working with antimony, which some have named "The Flamel Path" in his honor. We will examine this method in more detail later.

Basil Valentine 1450?

The most authoritative texts concerning the alchemical works with antimony come from Basil Valentine. His true identity is uncertain, although he claimed to be a Benedictine monk writing in the early 15th century.

In 1604, there appeared his work titled *The Triumphal Chariot of Antimony,* which described the preparation of many compounds from the ores of antimony, useful in medicine and alchemy. This text became almost the centerpiece of all future work with antimony and is referred to by many artists with the utmost esteem.

Basil Valentine was well aware of the poisonous nature of antimony in its raw state; in fact he called it "a very great and excellent poison". His methods detail the means of removing the toxic qualities from antimony in order to produce powerful medicinal agents for man and metals.

> Following the correct and true preparation of stibnite, there is no more poison to be found in it at all, for the antimonium must be completely transmuted by means of the spagyric art, and a medicine thus be produced from the poison.

Through the years, *The Triumphal Chariot of Antimony* was republished in many editions in several countries. The 1671 edition contains an extensive commentary by medical doctor Theodor Kerkring. Kerkring provides valuable insight on the various preparations of antimony given by Basil Valentine, and provides additional working knowledge to avoid pitfalls and dangers. Kerkring also provides his observations on the use of antimonial products in his own medical practice which serve to corroborate the amazing healing potential of antimony as claimed by Basil Valentine.

Valentine's descriptions for preparing the various products of antimony are so clear and concise that we have included exerpts form his works throughout this chapter to illustrate the processes.

Paracelsus 1493-1541

The work of Paracelsus was instrumental to the introduction of metallic derivatives in internal medicine. The theoretical aspects presented, and the style of writing in *The Triumphal Chariot of Antimony*, have led some to suspect that the true identity of its mysterious author was none other than Paracelsus himself, or at least the real author was heavily influenced by the doctrines of Paracelsus.

The writings of Paracelsus are filled with references to the preparation and use of antimony, though in a more obscure manner, and he is in full accord with Basil Valentine regarding the healing virtues of this material.

> In the same way that antimony refines gold, it also refines the body in the same form and manner; for within it is the essence, which lets nothing impure remain with what is pure. And no one who is experienced with all archidoxic writings, nor any

spagyrist, may fathom the strength and virtue of antimony.

Antimony lends itself to a surprising range of preparation methods and products. Spurred on by the works of Basil Valentine and Paracelsus, there was a great surge of interest in the medicinal use of antimony during the 16th century.

There were, of course, antimony preparations made without regard to alchemical principles which caused poisonings. Soon antimony had a bad reputation which prompted the Counsel of Paris to ban its use in medicine. This stand was also taken on by several other countries and the use of antimony for medicine was banned from 1566 to about 1650.

Eirenaeus Philalethes 1628-1665

Another great adept in the alchemical arts, who appeared in the 1600s, was the American alchemist George Starkey, writing under the name Eirenaeus Philalethes. His works include *The Metamorphosis of Metals, The Marrow of Alchemy,* and *An Open Entrance to the Closed Palace of the King.* A close study of these works is highly recommended to the practical artist. Philalethes followed a dry path very similar to that of Flamel. In *The Marrow of Alchemy,* Philalethes gives a description of the matter to be used:

> The substance which we first take in hand, is a mineral similar to Mercury, which a crude sulphur does bake in the Earth; and is called Saturn's Child, which indeed appears vile to sight, but is glorious within; it is sable coloured, with argent veins appearing intermixed in the body, whose sparkling line stains the connate sulphur; it is wholly volatile and unfixed, yet taken in this native crudity, it purgeth all the superfluity of Sol; it is of a venomous nature, and abused by many in a medicinal way; if its elements by Art are loosened, the inside appears very resplendent, which then flow in the fire like a metal, although there is nothing of a metallic kind more brittle.

Philalethes calls this "Our Dragon" and it is a description of stibnite, the sulfide ore of antimony. We will examine more of Philalethes method when we come to the Dry Way of working with antimony later on.

Isaac Newton 1642-1727

The great genius of Issac Newton was focused on the practice of alchemy for over thirty years. His notes indicate that he believed he was very close to perfecting the Stone of the Wise. Laboratory accidents involving metallic mercury vapors may have led to an early demise, a clear warning that "Dragons" are only approached with due precautions.

Newton was fascinated by the work on antimony and heavily influenced by the works of his contemporary, Eirenaeus Philalethes.

> In Antimony are Mercury (in the regulus), Sulfur (in the redness) and Salt (in the black earth which sinks to the bottom), which three, corrected, separated, and finally united together in the proper manner of Art so that fixation be obtained without poison, give an opportunity to the artificer to approach the Stone of Fire.
>
> Keynes Ms 64

MINEROLOGY

Antimony Ores

Antimony occurs in a wide variety of minerals in association with sulfur, oxygen, arsenic, copper, lead, silver and calcium. Many of these are somewhat rare, as is the native metal. The most common ores of commercial importance include the following:

Stibnite – this is antimony trisulfide, the number one source for antimony. When the ancient adepts speak of antimony, they are usually referring to stibnite. It forms in the orthorhombic crystal system and consists of 71.4% antimony with 28.6%

sulfur. Stibnite melts at the relatively low temperature of 550 deg C.

Senarmontite – antimony trioxide, also called "flowers of antimony". It generally occurs as octahedral crystals of the cubic system containing 83.3% antimony and 16.7% oxygen, melting at a temperature of 655 deg C.

Valentinite – another form of antimony trioxide named after the alchemist Basil Valentine. It occurs in rhombic crystal forms.

Kermesite – antimony oxysulfide, also known as antimony blende or red antimony ore. It is a mixture of two parts antimony trisulfide and one part antimony trioxide.

Cervantitie – Antimony tetraoxide, also called antimony ochre. It occurs as white or yellowish orthorhombic crystals with a pearly or greasy luster.

Chemistry of Antimony

Antimony rarely occurs in a pure state and so before use in alchemical works it must be cleaned and concentrated to some degree.

The most common impurities associated with antimony ores are arsenic, free sulfur, and mercury. Fortunately these are volatile materials and a simple roasting of the powdered ore will remove most of them. In this, proceed as described above in the section on preparing mineral ores.

Stibnite, being the principle ore of antimony, is also purified from its surrounding matrix rock by taking advantage of its relatively low melting point. This process is called "Liquation" and is often performed at the mine itself before shipping the ore to a refinery.

Liquation Process

This is a method for purifying and concentrating stibnite ores which contain 50% or more antimony sulfide. Lower grade ores are purified by different processes, one of which we will look at later.

The crushed ore is placed into special crucibles that have an opening at the bottom connected to a tube that runs out of the furnace, or to a cooler region at the bottom of the furnace. The furnace is heated using gas, coal, or even wood. As the crucible of ore heats up, the stibnite melts and flows out of the tube at the bottom into a catch vessel. Sometimes the catch vessel is filled with water so that as the molten stibnite hits the water, it cools and is shattered into finer particles. This makes grinding it later much easier. The residue of the ore remaining in the crucible consists mainly of the matrix rock, such as silica, which contained the stibnite, along with some of the stibnite that didn't melt out. It is saved for processing as a low grade ore.

The material that did melt out is often well over 95% pure antimony trisulfide and suitable for use in alchemical works; it forms especially nice glass and regulus.

We can use simple materials to perform this type of concentration easily.

For the crucible, an unglazed earthenware flowerpot with a hole at the bottom works well. You can find these at any garden supply in a range of sizes to fit your needs.

Use firebricks to construct a simple fire pit with a supporting metal can, having both ends cut out, placed in the center. Three or four nails poked in through the side of the can will hold it in place an inch or two from the furnace bottom. Below this can is placed a second larger can, such as a coffee can partly filled with water, to act as a catch vessel.

Fill the flowerpot with crushed stibnite, placing larger fragments at the bottom and finer portions at the top. You can cover the pot with a piece of steel screen to keep ashes out and also keep heat in. Heating is provided with charcoal packed all around the flowerpot as shown in the diagram below.

Collect all of the material that falls into the catch vessel as your purified stibnite, and be sure to save the residue remaining in the flowerpot, as it can be processed further by the Kermes method described below. Also, save the water from the catch vessel, as it can be used for preparing Vinegar of Antimony.

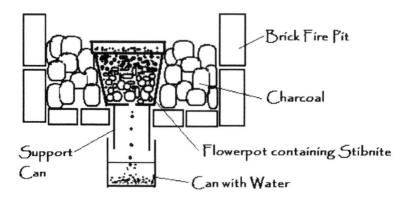

Brick Fire Pit

Charcoal

Support Can

Flowerpot containing Stibnite

Can with Water

Liquation of Stibnite

Kermes Mineral

Stibnite is easily soluble in strong alkaline solutions, forming alkali antimonates.

By taking advantage of this property it is possible to purify Stibnite, even low-grade ores, with a chemical process. The result of this purification is a red-brown powder called "Kermes Mineral", named after a dye of this color made from insects. Chemically it is known as Antimony Oxysulfide.

The preparation is easy but involves using a strong caustic solution and it produces a foul-smelling odor like rotten eggs (hydrogen sulfide), which is quite toxic to breathe, so this is best performed outside or in a fume hood.

Start by grinding the ore or the residue from the liquation process to a fine powder, then set it aside until we need it.

Now prepare a strong alkali solution by dissolving Lye (sodium hydroxide) into rainwater. A 20 to 30% solution works well; this will get very hot as the lye dissolves, so add it slowly to avoid boiling; also wear eye and hand protection. You can find lye in most hardware stores or even supermarkets; it is used to unclog plumbing. Be sure to buy the solid lye flakes or pellets and not the liquid mixtures which contain other soaps and surfactants.

Start adding the powdered stibnite to the still hot lye solution with stirring by a non-metal rod. The amount of ore added depends on its quality, but it is better to add it in excess to the

196

weight of lye used; we can do this alkaline leaching several times to pull out all of the antimony. The solution can even be heated to near boiling to hasten the dissolution of the stibnite.

After an hour of digestion, let the solution settle a bit, then filter it through a wad of glass wool. This solution is very caustic and will eat right through paper filters; you can find glass wool at aquarium suppliers.

The resulting solution will be of a deep golden yellow color. Into this solution, slowly pour in a 10 to 30% solution of acetic acid until the solution pH is 7 or neutral. Distilled white vinegar which has been concentrated by freezing and thawing once or twice, as we did for the acetate process, works well.

This is the smelly part mentioned earlier; a lot of hydrogen sulfide is released. Remember, this is toxic, so definitely be outside and upwind if not using a fumehood.

As the acetic acid is added, you will begin to see a red-brown solid form and fall to the bottom; this is the Kermes Mineral.

Allow the solids to settle, then decant the clear liquid from the top and save it aside. This liquid contains mostly sodium acetate which can be recovered for use in the acetate work. Its previous association with antimony makes it even more valuable.

The still moist Kermes is washed several times with rainwater by covering it with 10 to 20 times its volume of water and letting it settle, then decant and repeat.

Place the wet solid into a dish to dry. If the dried powder has a white crystalline crust, it means the sodium acetate was not washed out completely and you will have to repeat the water washing again to remove it.

The resulting red-brown powder, the Kermes, is now cleansed of many impurities that are associated with antimony ores, including arsenic, mercury and the alumina and silica matrix. Chemically the powder represents a complex mixture of antimony trisulfide and antimony trioxide.

Other alkalis will also work to dissolve the stibnite, such as potassium hydroxide, salt of tartar, even liquid ammonia. By altering the concentrations, and the order of mixing the acid and alkali solutions, the powder can be made to take on shades of canary yellow to brilliant orange to crimson red, as the particle size varies.

Kermes is much easier to calcine to a light oxide powder because of its greater purity.

Glass of Antimony

Another property of antimony is that it is a glass-former. Ancient glass and ceramic artifacts bear witness to this knowledge far back in time.

Preparation of the glass begins with the calcination of stibnite or kermes into the oxide. This conversion to the oxide need not be too rigorous, as a small proportion of the sulfide is needed to help form the glass.

If you melt pure antimony oxide you will see a beautiful orange liquid, but when it is cast and cooled it will revert to an opaque yellowish-white crystalline mass. The presence of antimony sulfide mixed into the antimony oxide will promote the formation of transparent glasses of intense red, yellow and orange shades.

The finely ground antimony oxide/sulfide is fused in a strong porcelain crucible at a temperature of about 700 to 1000 deg C, sometimes even up to 1300 deg C. It helps to add a small amount of the raw stibnite powder in doing this to obtain a deep ruby red glass.

Some use borax as a flux, but this often leads to problems in washing it out it later with water; in fact some claim that borax as well as aluminum lead to alchemical death of the subject and avoid their use at all costs.

Basil Valentine mentions the use of borax to form the glass, but this is a trap for the unsuspecting beginner. He also states clearly that glass made without any additions is the best for all uses.

When the crucible is ¾ full, and entirely molten, stick a thin iron rod into it and pull it out. Look at the glass adhering to the rod—if it is transparent, then it is ready; if it is cloudy, continue heating until it clarifies; however, don't overdo the fusion for too long as the material will be volatilizing throughout the process.

When the melt is ready, use tongs and quickly pour the liquid out into a wide flat dish of copper, then cover it with a lid because it often shatters and flies out of the dish on cooling.

Once cool, you will have the glass of antimony; it should be transparent and generally of a yellow to deep red tint, although by altering proportions and heating it is possible to obtain glasses of other colors, even green and blue.

The glass is held by many to be the preferred starting material for extracting the Sulfur of antimony.

Star Regulus

The term Regulus of Antimony has always been used to refer to the metallic antimony reduced from its ore. When it is well prepared and purified you will see a starry pattern on the surface of the metal. The metal itself is quite brittle and reduces to a powder fairly easily. Hidden in the Regulus is the spirit of antimony.

The most common way of producing the regulus begins with Stibnite or the Kermes mineral. Although these minerals can be directly smelted into the metallic regulus, they are generally mixed with a variety of fluxing and reducing agents which promote a greater yield and higher purity.

Isaac Newton recommended 2 parts stibnite, 1 part of iron filings, and 4 parts of burnt tartar. The mixture is fused in a crucible and allowed to cool slowly.

A slag or scoria forms at the surface and easily comes off the metal with a hammer blow. Save this scoria from the first reduction of the ore to metal; it contains the "Seed of Gold".

The metal or regulus at the bottom of the crucible may show some signs of starring, but generally requires additional purification by grinding and fusion with niter to bring out the stars.

The use of iron in the reduction produces iron sulfide by taking the sulfur from the stibnite, leaving the antimony free as metal which sinks to the bottom.

Other methods of obtaining the regulus include the use of niter in the mixture. For example, 12 parts stibnite, 5 parts iron filings, 6 parts niter, and 9 parts raw tartar. Even small iron nails can be used for the reduction in place of iron filings.

The inclusion of raw tartar is said to increase the yield of the "seed of gold" in the scoria during the first fusion.

The exact proportions will depend on the quality of the stibnite you begin with.

A word of caution: the mixtures with niter are essentially a form of gunpowder, which you will be placing into a very hot crucible. Add material slowly or you will soon find out why the ancients called this process "detonation".

The regulus, once obtained, is ground and mixed with twice its weight of niter, then fused again in a crucible to purify it. This purification may be repeated several times in order for the starry qualities to develop in the metal. This "Star Regulus of Antimony" is also called the "Martial Regulus" because of the iron used in its production.

Niter is a powerful oxidizer and the purification of the regulus with its aid must be done quickly or you will lose a significant portion of the metal as scoria. Some operators just sprinkle a thin layer of niter on the molten regulus for the purification step in order to reduce the amount of oxidation. Others suggest varying proportions of niter and salt of tartar.

Basil Valentine recommends the following procedure:

> Take of the best Hungarian Antimony, and crude Tartar equal parts, and of Salnitre half a part; grind them well together, and afterwards flux them in a Wind-furnace; pour out the flowing Matter into a Cone, and there let it cool; then you will find the Regulus, which thrice or oftner purge by Fire, with Tartar and Nitre, and it will be bright and white, shining like Cupellate Silver, which hath fulminated and overcome all its Lead.
>
> *Triumphal Chariot*

The final proportions of niter and salt of tartar, as well as the number of times the fusion must be repeated will depend on the quality of the stibnite used in the beginning. A little trial and error upon smaller quantities can save precious regulus from turning into scoria. Also, pouring the molten regulus into a hot iron mold and allowing it to slowly cool will enhance the crystallization into the Star formation.

Remember that many of these mixtures present an explosion hazard. The descriptive names for these processes used by the old adepts were pyrotechnics and detonation, so work slowly and with caution. Be sure to have adequate personal protection, including safety goggles or full-face shield and heavy gloves. Don't forget that you are dealing with a dragon.

Preparing the regulus of antimony is always an exciting undertaking which seems to create its own sacred space.

The illustration below is from *The Book of Lambsprinck*, written by a German adept around 1600. In it we see the "Scaly Dragon" (stibnite) being subdued by Mars (iron).

Alchemical Works

> Antimony is hermaphroditic, male and female, of both natures, Sulphur and, Mercury, fixed and volatile, the first-born of the metallic nature, middle substance between Mercury and metal, the only natural solvent and natural fire with which all things can be mixed, the Dragon and the devouring Lion, the solvent and the coagulant.
>
> Joseph Du Chesne
> *Treatise on Metallic Medicine,* 1641

In the modern periodic table of elements, antimony is listed under the group 5A elements along with nitrogen, phosphorus, arsenic, and bismuth. They all share similar properties in their chemical action. Alchemically speaking, these elements are held to be the activators or animators of matter.

Just as nitrogen and phosphorus are widely used in agriculture as fertilizers to induce heavy growth, antimony is often considered as the fertilizer of the mineral world. Antimony is held to bring in the necessary fire to accelerate metallic evolution.

In addition, as each of the metals is held to be under the rulership of a particular planet, antimony is said to ruled by the planet Earth. When the metal "dies", its spirit has nowhere to go but here again. Because of this, antimony is said to be immortal; its spirit is fixed to the sphere of the earth and can be transferred to other metals in order to reanimate them and awaken their generative power.

Vinegar of Antimony

One of the very important preparations from antimony is the Vinegar of Antimony, which is its pure fixed Mercury. It is called vinegar because it is prepared by a fermentation process and it tastes sour, like vinegar.

The fermentation itself takes some time, several months at the least, but the preparation is well worth the effort. The resulting "vinegar" is a menstruum which can extract a tincture from any of the minerals or metals and lend its life force to reanimate them. It also possesses remarkable healing virtues internally and externally. Some believe it is a specific against many forms of cancer.

Basil Valentine gives a very clear description of its preparation in his *Triumphal Chariot of Antimony* as follows.

> Melt the Minera of Antimony, and purify it, grind it to a Subtile Powder, this Matter put into a Round Glass, which is called a Phial, having a long Neck, pour upon it distilled Water, that the Vessel may be half full. Then having well closed the Vessel, set it to putrefy in Horse-dung, until the Mineral begins to wax hot, and cast out a Froth to the Superficies: then 'tis time to take it out; for that

is a Sign the Body is opened. This digested Matter put into Cucurbit, which well close, and extract the Water, which will have an acid Taste. When all the Water is come off, intend the Fire, and a Sublimate will ascend; this again grind with the Feces, and again pour on the same Water, and a second time abstract it, then it will be more Sharp. This Operation must be repeated, until the Water be made as Acid, as any other Sharp distilled Vinegar of Wine. But the Sublimate, the oftner the Operation is repeated, the more it is diminished. When you have obtained this Acid Vinegar, take fresh Minera as before and pour this Vinegar on it, so as it may stand above it three Fingers; put it into a Pelican, and digest it two days in Heat, then the Vinegar becomes red, and much more sharp then before. Cant this clean off, and distil it without addition in B.M. The Vinegar comes off white, and the Redness remains in the bottom, which extracted with Spirit of Wine is an excellent Medicine. Again rectify the Vinegar in B.M. that it may be freed from its Phlegm; lastly dissolve in its proper Salt, viz: in four ounces of it, to one ounce of the Salt, and force it strongly by Ashes; then the Vinegar becomes more sharp, and acquires greater Strength, and virtue.

You will notice that the process begins with the melting and purification of stibnite, which is the liquation process described above. Remember to use the catch basin water from this process where Basil directs you to add distilled water.

Theodor Kerkring, in his commentary to Valentine's *Triumphal Chariot*, adds some additional tips for success in the preparation:

> For six pounds of Antimony are required sixteen pounds of Distilled Water, and when (after Digestion) we would distil it, a certain manual Operation must be observed, on which depends

the Success of the whole Work almost. For the Alembeck must be so placed, as his Pipe or Beak may be covered with Water, which either must be put into the Recipient, or pass out by distilling into the same; otherwise the Spirit's of the Antimony will be lost, and more then half part of the same perish, or the Work require much more time for its perfection. When the whole Water hath passed over by Alembeck, the Fire is to be increased, and three Days, and as many Nights continued without intermission. Then let all cool, and the Sublimate, as he teaches, must again be mixed with the Antimony; this Labour for three Days and Nights must be re-assumed, and afterwards repeated to the third time. Then your Water will be acid, as common Vinegar. If you tinge this Vinegar with new Minera of Antimony, you will have a Tincture, which Basilius names his Balsome of Life, so often described, but never sufficiently commended. O, did Mortals know what Mysteries lie absconded in this Tincture, I question whether they would be desirous to set about any other Preparation of Antimony. All things are in this One. I have spoken, O Lover of Chymistry, do thou act.

It is important to calcine out any free sulfur prior to the digestion in order to prevent the formation of sulfuric acid. This should be a long, gentle calcination, slowly up to about 200 deg C maximum.

Save all of the antimony residues after the process, as they can still be used to prepare oils and the regulus.

Oil of Antimony

Also many Oils may be prepared of Antimony, some per se and without Addition, and many others by Addition. Yet they are not endued with the same Virtues, but each enjoys its own, according to the Diversity of its Preparation.

So Antimony, when prepared by the Addition of Water, assumes another nature and Complexion for operating, than when prepared by Fire only. And although every Preparation of it ought to be made by Fire, without which the Virtue of it cannot be manifested: yet consider, that the Addition of Earth gives it wholly another Nature, than the Addition of Water. So also when Antimony is sublimed in Fire through the Air, and further prepared, another Virtue, other Powers.

Triumphal Chariot

Fixed Red Oil

There are many ways to obtain an oil from antimony, and the properties of these oils vary with their method of preparation. One of the most important and valuable of these is the "Fixed Red Oil".

The preparation begins with the calcination of purified stibnite or kermes into the oxide form. The oxide is ground very fine and made into glass of antimony. This glass is powdered and then extracted with a strong solution of vinegar or still better, radical vinegar. Let the extraction continue for several weeks at about 40 deg C and agitate it once in awhile, especially in the first few days, or it will coalesce into a thick mass. After this time the solution will take on a golden to deep red color. Filter off the extract, and repeat with fresh vinegar.

Combine all of the extract and filter it into a distillation vessel. Gently distill the liquid until it becomes thickened, then add some water to dissolve the residue and continue the distillation.

Repeat this washing with water to remove as much of the acid as possible. This washing can also be performed using alcohol; in this case, ethyl acetate is formed and readily distills out so the washing is faster. The resulting residue will appear as a golden brown, gummy resin.

Place the resin into a suitably sized distillation vessel and proceed to distill as in the acetate work. Drops of a blood red oil will come over which are carefully collected by dissolving them into absolute alcohol. Rinse any of the oil adhering to the glassware out with alcohol and combine all of the liquid into a container. Seal and allow it to stand for several days, then decant the clear tinted extract for use. This "Fixed Tincture of Antimony" has powerful healing properties unrecognized by modern medicine.

Unfixed Oil

Antimony can also be extracted with a volatile solvent in order to produce an "Unfixed Tincture".

The most commonly used solvent for this is the so-called "Kerkring Menstruum" or "Philosophical Alcohol". This preparation is attributed to Kerkring who mentions it in his commentary to Basil Valentine's *Triumphal Chariot of Antimony.*

This is an example of a vegetable spirit being "magnetized" or "determined" to activity in the mineral realm by contact with prepared salts.

In this menstruum, we make use of the salt Ammonium Chloride (NH_4Cl) or Sal Ammoniac to sharpen the alcohol. Preparation of the menstruum begins with the sublimation of Sal Ammoniac.

This is easily done using CorningWare casseroles over electric or gas heat. After the first sublimation, the sublimed crystals will take on a pale yellow color; collect them and sublime again.

The sublimate will appear more yellow-orange and even reddish in areas; collect and sublime a third time. After this third sublimation, the crystals will appear very yellow orange to red-yellow and are ready for use. Store them in a glass container sealed from moisture.

Next, we need a very strong alcohol, 95% at least and preferably from red wine. The alcohol should be additionally dried by adding one or two ounces freshly dried salt of tartar (potassium carbonate) per liter. Allow this to digest at least for a day, then distil just prior to use.

When we have prepared these two ingredients, they are combined at the New Moon in the proportion of 4 parts Sal Ammoniac to 10 parts Alcohol. Seal them in a glass vessel and let digest at about 40 deg C for a month at least.

After digestion, the whole matter is gently distilled to near dryness. Collect the distillate and distill it again two more times. The final distillate will be the "Kerkring Menstruum"; seal it tightly in a glass vessel for use. Collect together all of the residues from the distillations and save them as well; they can be used to "charge" more alcohol several times.

One of the best solvents or menstruums to use for preparing the unfixed tincture of antimony is "The Acetone of the Wise" or "The White Wine Spirit" derived from the distillation of a metallic acetate. After isolating and rectifying the wine spirit, sharpen it with sal ammoniac just as was done with the Kerkring Menstruum.

We can use either of these prepared solvents to extract powdered antimony glass, calcined stibnite or calcined kermes.

This calcination should be done very slowly and carefully; at no time should globules of fused material be allowed to form.

A slow roasting at 90 deg C for a day is wise, then very gradually increase the heat with constant stirring until the matter begins to lighten. When the material is becoming lighter gray, indicating the sulfur is nearly vaporized out, the temperature can be increased. Grind to a fine powder, then continue the calcination to a light gray or even to whiteness. Always keep in mind that these vapors are very toxic, so have adequate ventilation or work outside.

Place the powder into a flask and cover it with one of the solvents, then seal it tightly. Allow this to digest at 40 deg C for a month or two at least. The menstruum will take on a golden color and slowly deepen to red-amber. Remember to shake it well once in a while.

Filter off the colored extract and save it aside. Now we have some options. First, we can slowly distill the extract to recover our menstruum and have a red oil remaining in the flask. The oil is dissolved into strong alcohol and allowed to stand several days before filtering for use. This is an "Unfixed Tincture from Antimony".

The second option provides an additional purification step for our antimony extract, providing that acetone was the menstruum. The extract we filtered off from the antimony is placed into a large, tall container. Into this, pour an equal

THE WAY OF THE CRUCIBLE

volume of a saturated solution of sodium bicarbonate (baking soda) in water. A brown precipitate will form and fall to the bottom. Decant off the liquid and save the precipitate. Wash it with water several times, allowing it to settle and decanting the water off each time. Finally collect the precipitate on a filter and let it dry gently.

The dried powder is extracted with strong alcohol which has been dried over potassium carbonate. The resulting tincture is filtered off for use and represents a more purified form of the Unfixed Tincture from Antimony.

In each case, fixed or unfixed oil, it is important to have little or no water present in the alcohol used for the final tincture extraction. This will prevent solution of any toxic antimony salts that may have carried through the process.

Also, the oil may be "sweetened" by gently distilling the tincture and then returning the distillate to the residue. This should be repeated several times. A small amount of "feces" will drop out with each cycle and the final tincture will mature.

Medicinally, the volatile or unfixed elixirs are warming, energizing, toning in their effects, while the fixed elixirs are cooling, and contracting. Unfixed tinctures are said to be more effective in acute illnesses, while Fixed tinctures are more useful for chronic diseases.

> Remedies that are unfixed heal unfixed diseases and the radically fixed nonvolatile ones expel fixed diseases which do not move the excrements through evacuation but through sweating and by other means.
>
> Isaac Newton, *Keynes Ms 64*

Another Oil of Antimony

> Sublime one part of Antimony, with a fourth part of Sal Armoniack, with subtile Fire. The Salt carries up the Sulphur of Antimony, red as Blood. Grind this Sublimate to a fine Powder, and if you took at first one pound of Antimony, grind with it again five ounces of Sal Armoniack, and Sublime

as before. The Sublimate dissolve in a moist place. Or otherwise, take the Sublimate, and edulcorate it from the Salt added, gently dry it, and you will have Sulphur, which burns like Common Sulphur, which is sold at the Apothecaries. From this Sulphur extract its Tincture with distilled Vinegar, and when you have abstracted the Vinegar by gentle Heat of B.M. and by a subtile Operation again distilled the remaining Powder, you will have (if in this Operation you erre not) a most Excellent Oil, grateful, Sweet, and pleasant in its use, without any Corrosiveness or peril.

The final powder mentioned above, which is distilled "by a subtile operation", is distilled as in the acetate process, by which you will obtain a red oil.

True Sulfur of Antimony

Take crude Hungarian Antimony, put that ground to a subtle Powder, into a Glass Cucurbit with a flat bottom: pour thereon the true Vinegar of Philosophers rendered more acid with its own Salt. Then set the Cucurbit firmly closed in Horse-dung, or B.M. to putrefy the Matter for forty Days, in which time the body resolves itself, and the Vinegar contracts a Colour red as Blood. Pour off the Vinegar, and pour on fresh, and do this so often, as until the Vinegar can no more be tinged. This being done, filter all the Vinegar through Paper, and again set it, put into a clean Glass firmly closed in Horse-dung, or B.M. as before, that it may putrefy for forty Days; in which time the Body again resolves itself, and the Matter in the Glass becomes as black as Calcanthum, or Shoemakers Ink. When you have this Sign, then true Solution is made, by which the further Separation of Elements is procured. Put this black matter into another Cucurbit, to which apply an Alembick, and distil off the Vinegar with Moderate Fire; then the Vinegar passeth out clear, and in the bottom a sordid matter remains; grind that to a subtle Powder, and edulcorate it with distilled Rain Water, then dry it with gentle heat, and put it in a Circulatory with a long Neck (the Circulatory must have three Cavities or Bellies, as if three Globes were set one above another, yet distinct or apart each from other, as Sublimatories, with their Aludel [or Head] are wont to be made, and it must have a long Neck like a Phial, (or Bolthead) and pour on it Spirit of Wine highly rectified, til it riseth three Fingers above the Matter, and having well closed the Vessel, set it in a moderate Heat for two Months. Then follows another new Extraction, and the Spirit of Wine becomes transparently red as a Ruby, or as was the first Extraction of the Vinegar, yea more fair. Pour off the Spirit of wine thus tinged, filter it through Paper, and put it into a Cucurbit (the black Matter which remains set aside, and separate from this Work; for it is not profitable therein) to which apply an Head and Receiver, and having firmly closed all Junctures, begin to distil in Ashes with moderate Fire: then the Spirit of wine carries over the Tincture of Antimony with it self, the Elements separate themselves each from other, and the Alembeck and Recipient seem to resemble the form of pure Gold transparent in Aspect. In the end some few Feces remain, and the Golden Colour in the Glass altogether fails. The red Matter, which in distilling passed over into the Receiver, put into a Circulatory for ten Days, and as many Nights. By that Circulation Separation is made; for the Oil thereby acquires Gravity, and separates itself to the bottom from the Spirit of Wine; and the Spirit of Wine is again Clear, as it was at first, and swims upon the Oil. Which admirable Separation is like a

Miracle in Nature: Separate this Oil from the Spirit of Wine by a Separatory.

This Oil is of a singular and incredible Sweetness, with which no other thing may be compared, it is grateful in the Use, and all Corrosiveness is separated from it. No man can by Cogitation judge, by Understanding comprehend, what incredible Effects, potent Powers, and profitable Virtues are in this Royal Oil. Therefore this Sulphur of Antimony, I have given no other Name, than my Balsam of Life.

Triumphal Chariot

Living Mercury Extracted from Antimony

Take the Regulus of Antimony, made in such manner, as I above taught, eight Parts. Salt of Humane Urine clarified and sublimed, one Part. Sal-Armoniack one Part, and one Part of Salt of Tartar. Mix all the Salts together in a Glass, and having poured on strong Wine Vinegar, lute it with the Luting of Sapience, and digest the Salts with the Vinegar for an entire Month in convenient Heat; afterward put all into a Cucurbit, and in Ashes distill off the Vinegar, that the Salts may remain dry. These dry Salts mix with three Parts of Venetian Earth, and by Retort distil the Mixture with strong Fire, and you will have a wonderful Spirit. This Spirit pour upon the aforesaid Regulus of Antimony reduced to a Powder, and set the whole in putrefaction for two Months. Then gently distil the Vinegar from it, and with what remains mix a fourfold weight of the filings of Steel, and with violent Fire distil by Retort: then the Spirit of Salt, which passeth out, carries over with itself the Mercury of Antimony in the Species of Fume. Wherefore in this Operation you must apply a great Recipient with a large quantity of Water in it,

211

so doing, the Spirit of Salt will be mixed with the Water, but the Mercury collected in the Bottom of the Glass into true living Mercury.

Triumphal Chariot

Salt of Antimony

Make a Regulus of Antimony, by Tartar and Salnitre, as I have above taught, grind this subtily, put it in a great round Glass, and place it in a moderate heat of Sand. This way the Antimony will be sublimed: whatsoever shall be sublimed, that dayly put down with a Feather, that at length it may remain in the Bottom, and there persist until nothing more of it can be sublimed, but the whole remains fixed in the Bottom. Then is your Regulus fixed, and precipitated per se. But consider, here is required a sufficient time, and repetition of the Labour often, before you can obtain that. This Red Precipitate take out, grind it to a subtle Powder, which spread upon a flat and clean Stone, set in a cold moist Place and there let it remain for six Months; at length the Precipitate begins to resolve it self into a red and pure Liquor, and the Feces or Earth is separated from it. The Salt of Antimony, I say, only resolves itself into Liquor, which filter, and put into a Cucurbit, that it may be condensed by extraction of the phlegm; and again set it in moist Place, then will yield you fair Crystals. Separate these from their Phlegm, and they will be pellucid, mixt with a red Colour; but when again purified become white. Then is made the true Salt of Antimony, as I have often prepared it.

Triumphal Chariot

The Fire Stone

The so-called "Fire Stone", or "Lapis Ignis", represents the summit of work on antimony itself. Utilizing the preparations of

212

antimony as detailed above, the Fire Stone is confected in a manner similar to the plant stone, where purified Body, Soul, and Spirit are reunited and congealed. Basil Valentine describes its preparation as follows.

> The Tincture of Antimony prepared fixed and solid, or the Stone of Fire (as I name it) is a certain pure, penetrative spiritual and fiery Essence, which is reduced into a coagulated Matter, like the Salamander, which in Fire is not consumed, but purified and conserved.

> Yet the Stone of Fire tingeth not universally, as the Stone of Philosophers, which is made of the Essence of Gold itself.

> PREPARATION Take in the Name of the Lord, of the Minera of Antimony, which grew after the Rising of the Sun, and Salt Nitre, of each equal Parts; grind them subtily and mix them; burn them together with a moderate Fire very artificially and warily; for in this the principal Part of the Work consists. Then you will have a matter inclining to Blackness. Of this matter make Glass, grind that Glass to a subtile Powder, and extract from it an high red Tincture with sharp distilled Vinegar, which is made of its proper Minera. Abstract the Vinegar in B.M. and a Powder will remain, which again extract with Spirit of Wine highly rectified, then some feces will be put down, and you will have a fair, red, sweet Extraction, which is of great use in Medicine. This is the pure Sulphur of Antimony, which must be separated as exactly as is possible.

> If of this Extraction you have one pound two ounces take of the Salt of Antimony, as I taught you to prepare it, four ounces and on them pour the Extraction, and circulate them, for a whole

213

Month at least, in a Vessel well closed, and the Salt will unite itself with the Extracted Sulphur. If Feces be put down, separate them, and again abstract the Spirit of Wine by B.M. The Powder which remains urge with vehement Fire, and not without admiration will come forth a varicoloured sweet Oil, grateful, pellucid and red. Rectify this Oil again in B.M. So that a fourth part of it may be distilled, and then it is prepared.

This Operation being completed, take living Mercury of Antimony, which I taught you how you should make, and pour upon it red Oil of Vitriol made upon Iron, and highly rectified. By Distillation in Sand remove the Phlegm from the Mercury; then you will have a precious Precipitate, in Colour such, as never was any more grateful to the Sight; and in Chronical Diseases and open Wounds, it may profitably be used for recovering the Pristine Sanity. For it vehemently dries up all Symptomatical Humors, whence Martial-Diseases proceed; in which the Spirit of the Oil, which remains with the Mercury, and conjoins and unites itself thereto, powerfully helps.

Take this precipitate, and of the Superior Sweet Oil of Antimony, equal parts, pour these together into a Phial, which well closed set in convenient heat, and the Precipitate will in time resolve and fix itself in the Oil. Also the Phlegm by the Fire will be consumed, and what remains become a Red, dry, fixed and fluid Powder which will not in the least give forth from itself any Fume.

Now my Follower, and Disciple of Arcanums, I will speak after a Prophetic manner. When you have brought your Philosophic Studies (in the

Method by me prescribed) to this end, you have the Medicine of Men and Metals.

Triumphal Chariot

Another method of preparing the Fire Stone, advocated by contemporary artists, begins with the preparation of Kermes as detailed above.

The Kermes is rinsed with rainwater until neutral. Be careful in this rinsing that the water does not become acidic, as that is a sign that some of the vinegar of antimony is beginning to come out and be lost with the rinse water.

Decant as much of the water from the Kermes as you can and then extract the still moist Kermes with a strong vinegar solution, at least 30 to 60% acetic acid. Let this digest at about 60 deg C, until the solution becomes golden yellow to orange in color.

Decant the solution and repeat the extraction with fresh vinegar. Combine all of the extracts and filter until clear. Now gently concentrate the solution by distillation or simple evaporation in a dish and allow crystals to form.

The collected crystals are then dry distilled as in the acetate path using a suitably sized apparatus. The distillation train shown below is suitable for this part of the process.

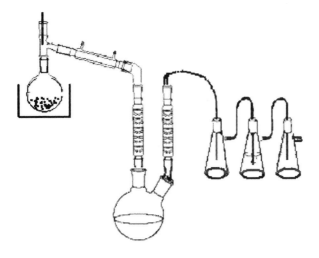

The distilling flask on the left can be heated using a sandbath. The central flask of the bubbler train is partially filled with alcohol; the other two flasks are left empty and act as guards against sudden pressure changes.

As the distillation proceeds, a clear watery "phlegm" will come over. Save this aside and continue heating. Soon the white vapor and red oil, as seen in other acetate distillations, will come over.

The red oil is the Sulfur of Antimony. The alcohol in the bubbler will become saturated with the Mercury of Antimony.

This Mercury of Antimony is not fixed as it is in the preparation of Vinegar of Antimony, so the alcohol solution must be kept cold or the Spirit will escape. Since it does not stay around for very long, it is a good idea to use it as a menstruum soon after preparation. It will extract the Sulfur from most prepared minerals and metals.

Collect the red oil and save it aside in a tightly sealed vial until we are ready to use it later. It has great medicinal virtues on its own, but our aim here is to unite it with a prepared body and form the Fire Stone.

The body we need for this is hidden within the solid residue remaining in the distillation flask. This black residue is called the "Black Lion".

Remove the black residue from the flask and grind it very fine, then place it into a clean, unglazed crucible. The crucible is then heated in a furnace to about 1000 deg C. The Black Lion will become lighter and possibly seep through the crucible, so it is usually set on an unglazed tile or pottery shard with a layer of Kaolin (from a pottery supplier).

Once cool, the crucible is placed into a strong glass container and covered with water. If the crucible leaked out onto the tile and kaolin, these are added to the container of water as well.

Heat the water to a boil so that the contents of the crucible will be leached out and dissolved into the water. Filter the liquid and gently evaporate until crystals form.

Collect these crystals and keep them dry and out of air contact, as they will readily deliquesce. This is the Salt of Antimony. The crystals are called the "Magnets of the Philosophers"; they have the property of fixing the Philosophical Mercury.

Recharge the distillation apparatus with a fresh load of "Kermes Acetate" and instead of placing alcohol into the bubbler train, place the dried crystals you collected.

Now repeat the acetate distillation to collect more red oil. In addition, the crystals in the bubbler train will become saturated with the Mercury of antimony and it will be fixed by the Salt.

Place the crystals into a vial; this is now the Salt united with the Mercury of antimony. Use the collected red oil to saturate the Salt and Mercury, then seal the vial and digest at 40 deg C initially. After about a week, increase the temperature to 60 deg C and continue the digestion. Similar to the plant stone, the matter will congeal into a translucent red solid, the Fire Stone.

Via Humida with Antimony

The alchemist Artephius described a process for confecting the Philosopher's Stone using Antimony around 1150 CE. It became known as the Wet Way or Via Humida.

The method involved two processes he called the First and Second Perfections. In the First Perfection, pure antimony metal is formed from stibnite when iron substitutes for the antimony in a liquid mixture. Iron sulfide forms a dull matte that is fused to the Regulus. During the process, Antimony Trichloride is also formed and distilled from the reaction mixture.

The reaction takes place through the agency of Ignis Innaturalis or Secret Fire, which is prepared from urine enhanced by the addition of potassium chloride, which was called Muriate of Potash.

The powdered stibnite and iron are mixed to form the "Compost", which is moistened with the secret fire and placed into a distillation train. The proportions are 3 parts pulverized Stibnite; 1 part Iron powder; and 4 parts Secret Fire (urine saturated with Potassium Chloride).

The neck of the flask must be kept cold so the vapor of the "Mercury" will condense.

Gently heat the flask in a sand bath, at a temperature of 40 deg C held constant for about three months.

There is a slow reaction as the Compost absorbs the liquid in the first few days. It swells, and sweats with a dark and rusty appearance. This is the Rule of Saturn of the First Perfection.

At the end of this digestion, a shiny star appears on the surface. This is the Regulus, which is pure metallic antimony. Antimony Trichloride vapors start to rise only after the Regulus forms.

The vapor of Antimony Trichloride forms as a white "metallic, volatile humidity", called the "Mercury of the Wise", the "white wife" or White Goddess of the First Perfection. This vapor, when distilled, becomes a clear shining liquid, the "water that does not wet the hands".

The condensed Antimony Trichloride is kept sealed in the receiving flask. It will fume slightly in air and is very corrosive, irritating and hygroscopic. This Mercury of Antimony Trichloride becomes "our vinegre" or Secret Sophic Fire in the Second Perfection.

Remove the matter from the distilling flask and separate the regulus from the black matte of iron sulfide. This is the "First Stone". Pulverize the Regulus in a mortar and pestle and save it aside. In the Second Perfection, this Regulus is called Our Mercury.

The Second Perfection is performed in a "Philosophical Egg", which is a flask having a long neck sealed with a ground glass stopper or silicone rubber stopper. The "egg" is filled to one quarter of its volume with the "Azoth", which is analogous to the Compost of the First Perfection. It is made from a mixture of Gold Leaf, Regulus, and the Secret or Sophic Fire of Antimony Trichloride produced in the First Perfection.

The ratio is Gold, 3 parts; Regulus, 1 part; and 4 parts of the Secret or Sophic Fire.

A period called the Rule of Saturn takes about 2 months for the matter to form a black mass. Keep the heat at about 40 deg C during this period. The matter will sweat and circulate in the flask, passing through the "Solve et Coagula" phase.

With time, the moist slime changes to a dry coagulated substance. The so-called Lyon or Toad's exhalations are the vapors seen in the neck of the flask, often compared to two fighting dragons, one with wings and the other without. Seven or eight successive volatilizations are necessary. Some alchemists call the volatilizations Eagles. These Eagles feed on the Lyon and drip back down on the Earth as dead eagles.

By the end of this period, the matter swells and becomes dark. A phase of calcination begins, which to leads to the black or Nigredo phase. The fumes cease. The Earth bursts and reduces into a black powder called the Raven. The Nigredo brings the end of the Rule of Saturn.

The Rule of Jupiter now begins. It takes about two more months before the "Peacock's Tail" appears. Slowly increase the temperature to 45 deg C. Then a water washes the Nigredo and beautiful colors appear, Alchemists called this the Peacock's Tail. It signals the end of Jupiter's Rule.

The Rule of Luna now begins. It takes about 28 days in Luna's Rule to achieve the Second Stone, which is when the matter turns white. This period was called the "albedo", or whitening, or the white dove. It floats on the water like a cream.

The Rule of Venus is 42 days more with heat increased to about 70 deg C. The King (gold) unites with the Queen (albedo). The Queen is also called Philosophical Mercury, rosa alba (the white rose). In uniting, the king and queen remove impurities from each other. The King and Queen produce the Green Lion.

The Rule of Mars is 42 days. During this time azure, gray and citrine overcome the Green Lion, which finally changes to a red color. This red matter is now called "Sulphur". The red Sulphur fixes the white Mercury and they dissolve into a liquid state.

The Rule of Sol is another 42 days, with the temperature increased to between 100 deg C and 121 deg C. In a short time, a crimson or Tyrian purple color forms as a sparkling oil. Alchemists call it the Red Rose; the perfect fixation and perfection; the Red Poppy of the Rock; and the precious tincture. This is at last the Philosopher's Stone or "Third Stone". When powder forms from the solution, the alchemist slowly increases the temperature to 150 deg C.

The red powder is a form of gold chloride. Traditionally, the blood red powder is the completed Philosopher's Stone.

For transmutation, melt metallic gold in a crucible, then add the red powder of the Philosopher's Stone in equal part to the crucible and pour the mixture into an ingot where it cools.

Now break away part of the ingot and pour ten parts of heated Mercury onto it. Melt the new mixture, which becomes tinged into pure gold.

Via Sicca with Antimony

The alchemical text *Coelum Philosophorum* (The Heaven of the Philosophers), by the German alchemist Phillipus Ulstadius around 1536, presents detailed instructions for working with many mineral and metallic substances. Be careful not to confuse this text with a work by Paracelsus with the same name. Ulstadius sheds a great deal of light on the subject of antimony and particularly the use of antimony regulus.

By far the greatest interest surrounding the regulus is its fixed spirit or Mercury, which can be transferred to other metals in order to reanimate them and awaken their generative power. This path leads to confecting the Philosopher's Stone and is referred to as Via Sicca or the Dry Way with antimony. With some operative variations, this is the path of Flamel, Philalethes and Newton. It is said to be one of the fastest and most reliable ways to make the Stone of the Wise, but filled with many dangers.

To begin, stibnite must be purified by the liquation process as described above. The purified stibnite is powdered and dried, then mixed with an equal amount of fluxing powder consisting of 2 parts salt of tartar and 1 part niter. Grind these together carefully and dry.

Next, a crucible is heated red hot and the dry powder mixture is slowly added spoonful by spoonful with a cover placed on top with each new addition. A scoria will form at the top and regulus will fall to the bottom of the crucible. Pour the melt into a hot cone-shaped iron mold and tap the sides to help the regulus fall to the bottom. Once cool, the regulus can be separated from the scoria easily by a hammer blow. Save the scoria aside; it contains the "seed of gold".

To purify the regulus, it is melted in a clean crucible and niter is sprinkled on top to form a thin molten layer. Keep this in flux for about 15 minutes, then pour it again into the iron cone. Repeat this purification two more times at least.

The next step is to form the Martial Regulus. The recommended proportions are 1 part regulus to $1/10^{th}$ part iron or steel nails. Iron is said to have a volatile magnetic quality and steel, a fixed magnetic quality. Steel is preferred here, but not stainless steel. Heat the nails red hot in a crucible, then slowly add the regulus as a powder. Keep in flux for 15 minutes, then pour into the iron cone which has been preheated. When cool, break the regulus away from the remaining undissolved nails. Purify the regulus as described above until you see the starry crystal surface.

Now melt your "Stellate Martial Regulus" in a clean crucible and slowly add pure silver metal filings until they stop melting, signaling it is saturated. This is now a Lunar Martial Regulus.

This is sufficient if you plan to make the White Stone which can transmute metals to silver. If you wish to make the Red Stone which transmutes metals to gold, you will need to add a small amount of copper.

Melt the Lunar Martial Regulus and slowly add pure copper metal powder until it refuses to melt anymore. It won't take very much. Cast the metal out and you will have a Lunar Venusian Martial Regulus with a beautiful violet sheen. Save this regulus aside for the amalgamation phase of the work.

> The Stone of the Philosophers is nothing more than a subtilized, exalted and seminal gold.
>
> Mercury is the agent which penetrates and effects such a subtilization and exaltation. But it must be prepared first for use in this Art.
>
> *Coelum Philosophorum*

Before we can use the prepared Lunar Venusian Regulus to "animate" metallic mercury, we need to purify mercury with some of the Martial Regulus.

Ideally we should use a very pure Native Cinnabar for this process, but if that is not available, we can use high purity metallic mercury (triple distilled) from the chemical supply store. It is a good idea to purify it in the old ways prior to use. It's not so much aimed at getting the mercury any more pure as at exalting it alchemically, opening the body to receive new life.

For this, take your mercury and wash it well with rainwater, then squeeze it through a piece of chamois or other thin pliable leather.

Now cover the mercury with dried and powdered sea salt. Mix these two completely using a mortar and pestle. Depending on the impurities in the mercury, the salt will darken even to blackness. Wash this dirty salt out using rainwater and repeat the salt washing once or twice. The mercury will retain some of the subtle essence of the salt which is important for the rest of the processes.

Put the mercury into a mortar and add an equal amount of sea salt. Saturate this mixture with strong distilled vinegar (about 10% acetic acid) and begin vigorously mixing the mass for about 10 minutes. Now wash this salt out with rainwater until the mercury appears shiny and bright. Press it through a chamois; there should be no residue left in the chamois. The mercury should leave no trail when rolled across a smooth surface and should be without any appearance of scum at the surface.

Mix the mercury with an equal amount of native sulfur in a mortar, grinding it together well. The matter will turn black during this process, forming mercury sulfide. Examine the matter closely with a magnifier. There should be no tiny globules of mercury remaining. Grind with additional sulfur if there are. The resulting black mass is a crude form of cinnabar, which is the sulfide ore of mercury. We can proceed with the black variety of cinnabar or we can improve this matter by sublimation to obtain the beautiful orange variety of cinnabar which is best. This is a difficult and dangerous process which requires careful control of the heating.

Now mix 1 part of the cinnabar with ½ part of powdered martial regulus and grind them together very well in a mortar. Place the powder into a strong glass retort with the distilling arm immersed in a container of water.

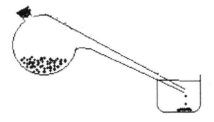

Begin the distillation slowly at first, then gradually raise the heat until metallic mercury distills over. The temperature will be about 320 to 380 deg C as the mercury distills. When the distillation is complete, remove the receiver first, before removing the heat, or the cooling retort may draw the water into itself and explode.

The black residue remaining in the retort is mostly antimony sulfide which can be used to make more of the Martial Regulus.

The mercury which distilled over is ground together with Native Sulfur until a homogeneous black powder of cinnabar is formed. Again, it is best to sublimate this black form of cinnabar into the red orange variety, but either form will work.

The process of mixing the cinnabar with Martial Regulus and distilling is repeated at least three times. The metallic mercury from the final distillation is now purified and prepared for the amalgamation phase.

Take 1 part of the Lunar Martial Regulus for the White Work, or 1 part of the Lunar Venusian Martial Regulus for the Red Work, as prepared earlier. Powder the regulus and add 4 parts of the purified mercury in small increments, grinding the matter very well in an iron mortar with each addition. Continue grinding the mass for 10 to 20 hours total. At the end of this grinding, add rainwater and continue to grind until the water is black. Pour out this dirty water and add fresh rainwater. Continue this water wash until all of the blackness is out and the resulting amalgam appears bright and clean.

Now place the amalgam into a strong retort as above or into a distilling flask set up as shown below.

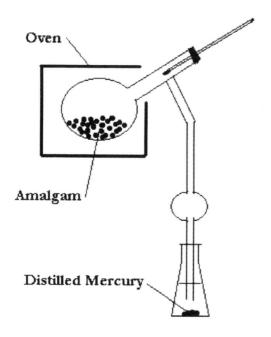

Oven

Amalgam

Distilled Mercury

The oven should be able to reach about 400 deg C. The bulb in the distilling arm prevents water from being drawn into the hot flask if there is a sudden pressure change. The receiver is partially filled with water to condense the mercury vapor distilling over.

Distill the amalgam and collect the mercury from the receiver. Wash the mercury several times with water and then dry it with a paper towel.

Remove the black silver and regulus residue from the distillation flask and weigh it, then grind it to powder. Melt half its weight of fresh regulus in a crucible, then slowly add the powder until it melts. Cast out the melt and powder it.

Use the distilled mercury to form the amalgam again with this fresh regulus as you did above. Now repeat the distillation.

The amalgamation and distillation process is repeated 7 to 10 times and is often referred to as "letting the eagles fly". The regulus of antimony won't amalgamate very easily with mercury metal, so the silver is added to absorb the fixed spirit of antimony and transfer it to the mercury.

The silver, then often called the "Doves of Diana", acts as a medium to transfer the life force of the antimony into the mercury. After the distillation, the residue of silver which remains in the retort, called now the "Dead Doves of Diana", is cleaned and used again each time with the addition of fresh regulus. With each cycle of amalgamation and distillation, the mercury metal becomes more enlivened and ultimately is called "Animated Mercury", or "Sophic Mercury", containing the generative power of the metallic realm.

This is the "Philosophical Mercury" of the Dry Way. Thus common mercury becomes reanimated by the fire of antimony and the principles of life hidden in the iron and copper. It is the fertile field wherein the seed of metals is sown.

For the White Work, purified silver is "seeded" into this animated mercury; for the Red Work, we use purified gold.

The purification of the precious metals begins with the preparation of sea salt. The salt is fused in a crucible at about 800 deg C, then dissolved in rainwater. Filter the solution into a distillation train and distill out 2/3 of the water. Save this distilled water aside, and pour the remaining salt concentrate into a dish.

Crystals will form in about a day, which are then collected. Dissolve the crystals in the distilled water you collected. Filter the solution into a dish and allow it to crystallize. Collect and gently dry the crystals. This is now purified salt to be used in preparing the gold and silver metals.

For the White Work, silver metal is dissolved in Aqua Fortis or nitric acid. Filter the solution through glass wool and dilute it with about ten times as much water. Now pour the solution into a clean dish made of copper. Some of the copper will dissolve and fine crystals of silver metal will form on the bottom of the dish. Wash this fine silver powder several times with rainwater, then dry it.

Grind together one part of the dried silver powder with six parts of mercury which has been purified with vinegar and salt. Press the resulting amalgam through a leather chamois to squeeze out the excess mercury.

Take the ball of amalgam remaining in the chamois and grind it together with the previously purified salt. Add the salt in small amounts and continue grinding until you cannot see the metallic amalgam in the powder. Place this powder into a distillation train and distill out the mercury. The residue remaining in the distillation flask is rinsed several times to dissolve and remove the salt. A very fine "silver calx" will remain.

Repeat the process of amalgamation with mercury, grinding with purified salt, distillation of the mercury, and washing the silver calx, two more times. The resulting fine silver calx will be suitable for seeding into the animated mercury and preparing the White Stone.

The ancient artists thought of this process as a type of calcination of the precious metals, wherein they are reduced to a fine ash like consistency, and hence the name "calx" of the metal.

The calcination of gold, for the Red Work, follows a similar path, and several methods are described later in the chapter called "The Book of Gold".

Now that we have prepared the metal calx and the Animated Mercury, we are ready to proceed with the next phase of the work. This part of the work consists of planting our precious metal "seed" into the Sophic Mercury.

Take one part of silver calx (for the White Work) or one part of gold calx (for the Red Work) and grind it together with two parts of the Sophic Mercury for about ten hours. The resulting amalgam is then washed with rainwater until no more blackness comes out in the water and the amalgam appears clean and bright.

Place the amalgam into a dish and cover it with a paper to keep dust out. The amalgam will become dry and firm up enough so that it can be shaped. Cut pieces of the amalgam off and form them into small balls about the size of a pea. Set the balls into a dish and let dry gently in the sun.

Place the dried amalgam balls into a heavy walled digestion flask with a ground glass stopper. The flask should be no more than a third full. Cover the flask with a piece of paper to keep dust out, then place it into a sandbath at a gentle heat of about 50 deg C for a day.

When you perceive that all of the moisture has gone out, seal the flask tightly with the ground glass stopper. Continue the digestion at a gentle heat.

After about a week, the balls of amalgam will come together as a mass which begins to swell and puff up. Gradually raise the temperature and continue digesting the flask.

The matter will become very dark but soon will begin to look lighter and finally take on a reddish hue. Increase the heat still higher and the red will become more pronounced after about three months. After six months of continuous digestion, the matter will be of a uniform red color.

This is the "Philosopher's Gold" and represents "the true radical and central calcination of gold". These are "the ashes of the philosophers, wherein the Royal Diadem is hidden" and is "the nearest matter of the Stone, out of which the Great Work may be made".

This "Sophic Gold" provides the foundation for confecting the Red Stone of the Philosophers, but first we must augment or multiply its virtue and quantity.

Place 3 parts of the Sophic Gold into a clean digestion flask as above and heat it until you can barely stand to touch it for very long. Using a heated glass funnel, you must add 1 part of preheated Sophic mercury in small portions. Add about the quantity of a pea each time and allow about half an hour to pass before the next addition. Each time the moisture should be allowed to vapor away.

When the full 1 part of Sophic Mercury has been added and all of the moisture has disappeared, seal the flask with the ground glass stopper. Continue the digestion in a sandbath for four weeks, gradually increasing the heat during this time. At no point should the temperature rise high enough to sublime or distill the mercury within the flask. After four weeks of digestion, the added mercury will become fixed by the Sophic Gold.

Increase the heat until the flask is almost glowing red hot and the matter will fuse without fuming. Allow the flask to cool slowly, then break it free of the digestion flask.

You can repeat this process of multiplication as many times as you like and thus have a constant supply of Sophic Gold.

Now take 1 part of the Sophic Gold reduced to powder and place it into a clean digestion flask. Add two parts of Sophic Mercury all at once, then gently warm the flask to exhale all of the moisture. Seal the flask and continue the digestion until the powder and mercury form a fluid amalgam. Pour the amalgam into a clean glass mortar and begin gently grinding it with rainwater until no more blackness enters the water. When the wash water comes off clear, pour the amalgam into a dish and let it dry in the sun.

Place the dried amalgam into a clean digestion flask and gently heat to remove all moisture before sealing. The flask is digested in a sandbath at about 40 deg C and left undisturbed.

The matter will ferment, swell and bubble, then turn black like pitch. This is called "The Regimen of Saturn". Keep up this gentle digestion for 40 or 50 days and the mercury will begin to circulate. Within several months, the matter will become lighter, and finally a pure white powder will form. This is the "White Sulfur of the Philosophers".

The final phase of this process is the preparation of the White and Red Stones from this White Sulfur. For the White Stone we use pure silver metal and for the Red Stone we use Pure Gold. In each case the method is the same, so we will only describe here the preparation of the Red Stone.

Place the White Sulfur into a digestion flask and add small portions of Sophic Mercury until the matter is a soft dark mass. Warm the flask to exhale all moisture, then seal it tightly.

Digest with a gentle heat in a sandbath for 30 days and the darkness will begin to lighten. Continue the digestion for another 30 days and the matter will take on a greenish hue, then a blue green, similar to a peacock's tail, which is what this phase of the work is called. As the digestion continues, the matter will become increasingly yellow, then reddish. At last, the entire mass will become a beautiful red powder which is called the "Red Sulfur of the Philosophers".

Take 1 part of this Red Sulfur and encase it in a shell of beeswax. Now melt 2 or 3 parts of pure gold in a crucible and cast the Red Sulfur wax ball into it. Cover the crucible and let remain molten for half an hour. Once cool, break open the crucible and you will find a glassy material which you must reduce into powder.

Place 1 part of this powder into a digestion flask and add 2 parts of the Sophic Mercury. Exhale the moisture out, then seal the flask. Digest the flask with a gentle heat as before. The matter will become dark at first but will turn red much quicker than the first time.

Repeat the digestion with fresh Sophic Mercury several times. The matter will turn red sooner each time. At the end of this process, you will have "The Red Stone of the Philosophers", "The Medicine of Metals", which can transmute impure metals into gold.

The process of transmutation of metals into gold is called Projection. Making a judgment of how much metal your stone will transmute requires a little experimentation.

Take about 4 parts of pure gold and melt it in a clean crucible. Cast in 1 part of your Red Stone wrapped in beeswax and keep the matter in flux for half an hour. When cool, break out the resulting red glassy mass from the crucible and powder it.

Now heat 100 parts common mercury in a crucible and cast in 1 part of the Red Stone wrapped in beeswax. Keep the matter in flux for half an hour, then cast it out into an ingot mold. If the metal is still brittle, you must remelt it with more mercury. If all of the metal did not change into gold, you must add more of the Stone. After a few trials, you will be able to judge the proportions in future transmutations.

In order to multiply the Stone in quantity and virtue, melt 1 part of it with 2 parts pure gold. Powder the resulting glassy mass and place it into a digestion flask. Add Sophic mercury until it becomes a soft mass, then seal the flask. Digest at a gentle heat and it will become fixed into the Red Stone in a short time.

With slight modifications, this is the same process for confecting the Philosopher's Stone advocated by Flamel and Philalethes. Some modern artists even refer to this as the Flamel Work because his was one of the earliest descriptions of it.

The healing and rejuvenative virtues hidden in Antimony certainly earn it a place as one of the great Rasayanas of Western alchemy.

Analytical Data

Appendix IIB and IIC contain a selection of analytical data collected on several types of oil obtained from antimony.

CHAPTER SEVENTEEN

The Book of Gold

Gold is probably the first metal to come into mankind's recognition. It's one of few metals that occur in nature as an actual metallic body, which is called "native gold"; it's never been smelted. Copper is often found in native form but is more difficult to recognize as it is susceptible to corrosion, as are most other metals, but not gold.

Bury gold in the ground or under the ocean for a thousand years and when you dig it up, it still glows warm like solidified sunlight. In fact, from ancient times, gold has been described as congealed sunlight.

CHEMISTRY OF GOLD

Recovery of gold from old jewelry
We can recover gold from old jewelry in a fairly easy way and obtain some useful experience with the chemistry of gold. Gather together all of the old gold jewelry you want to work with. This can be broken chains, bracelets, old rings, whatever.

The overall weight of gold should not exceed twenty-five percent of the total metal. This is estimated from the carat weight of the gold item and its total weight. If there is too much gold, it must be diluted to twenty-five percent or less of the total weight by adding silver metal. This is called quartering or inquartation and is done to assist the subsequent acid treatments.

Gold lends a certain amount of protection to lesser metals associated with it and they become more resistant to acids. By diluting the gold, the other metals become susceptible to acid attacks. Don't worry, you can recover the added silver later along with any silver from the jewelry.

The entire mass of collected jewels and silver (as needed) is melted in a crucible, then rapidly quenched by pouring it into a container of water. This will turn all the metal into a fine shot of homogeneous composition. Decant the water and collect the metal residue into a glass container.

Slowly add concentrated hydrochloric acid to cover the metal. This removes metals like tin, iron, zinc, and aluminum as soluble chlorides. Decant the liquid and wash the solids with a little water.

The solid residue is now treated by the slow addition of nitric acid. This will dissolve the silver, and any other metal still associated with the gold. The gold remains unaffected and falls to the bottom as a fine brown powder. After settling, decant the acid solution and save it aside to recover the silver. The nitric acid treatment can be performed a second time on the gold residue and heated; save the acid solution with the first.

Wash the gold residue with water a few times, then dissolve it in aqua regia, which is three parts hydrochloric acid mixed with one part nitric acid. Allow the deep golden yellow solution to stand awhile because a small amount of impurities may settle in the form of gray or white silica and alumina which were occluded in the jewelry casting processes. The clear solution can be evaporated down with additions of hydrochloric acid to remove the nitric acid. Evaporate the final liquid down to obtain tetrachloroauric acid ($H\{AuCl4\}$) crystals, also called gold chloride, for alchemical works.

In order to obtain metallic gold, you must decant the clear golden acid solution into a clean vessel, then slowly add a solution of ferrous sulfate (green vitriol) in water. The gold will immediately fall to the bottom in a very pure form as a golden brown heavy solid. Let it settle, then add more vitriol solution until no more gold will fall out.

Wash the gold powder with water and let it dry. This finely divided gold, often called a gold calx, can be used as is for alchemical works or it can be placed into a crucible, melted and cast into an ingot.

In order to recover the silver, we can add metallic copper or iron to the nitric acid solution we saved earlier. Very quickly you will see crystals of metallic silver growing on the added metal. These crystals can be shaken off, allowed to settle and then washed with water and dried.

As an alternate method, or to insure we have recovered all of the silver from the solution remaining from the above method, treat the nitric acid solution with a solution of common salt in water.

A white precipitate of silver chloride forms and settles to the bottom; in fact if you stir the solution for a while, the particles will agglomerate and fall sooner when you let it settle. Decant the solution and discard or save for another work; it mostly contains copper. Wash the white precipitate with water a few times to get all the salt out, then let it dry in a dark place. It will rapidly turn from pure white to purple or gray in sunlight.

The dried white silver chloride can be used for alchemical work, where it has been named "horn silver" and silver calx.

If you want to regain the silver as a metal, you can mix the silver chloride with dilute sulfuric acid and add zinc metal to saturation. The residue is washed, dried, then melted in a crucible and cast as an ingot. The casing of an old flashlight battery will provide you with a pretty good quality zinc metal for use in this process.

An alternative method begins with the freshly precipitated silver chloride which has been washed and decanted into a thick slurry. Add solid sodium hydroxide (lye) to cover. Caution! This gets very hot very fast and tends to splatter if you add the lye too quickly. Protect your eyes and skin. When the reaction subsides, sprinkle in a sugar, like dextrose (grape sugar), or fructose, or even corn syrup, to reduce the silver into metal. Table sugar (sucrose) will work but not as effectively as the simple sugars. The mass will darken and the silver falls to the bottom. When addition of sugar no longer causes a reaction, add water to wash the silver several times. It settles quickly to the bottom.

This is a finely divided silver which can be used as is, or melted into an ingot.

Process of Fire Assaying

Fire assaying is an ancient method of determining the amounts of gold and silver in a mineral ore or a metallic article

233

through a process of fusion with selected salts. This method, perhaps the most ancient analytical technique, is still in use today for determining gold and silver values in the mining industry.

There is evidence of fire assaying being practiced in the ancient mediterranean area as early as 2600 BCE. In the development of alchemy, the methods of the fire assay provided a means of unlocking the mineral world.

Even today, fire assaying is considered as much an art as it is a science; but then that makes sense when you realize that the methods sprang up from the "Dry Ways" of alchemical work, and the fire assay has grown up with alchemy in close association through the centuries.

The process of fire assaying can provide us with some valuable insight to alchemical works as an example of dry working and the action of various flux ingredients used in mineral works. In addition, there are phrases commonly used in fire assay which you will come upon in the study of alchemical texts, such as "trial by the fire" and "cupellation", so here you will get an understanding of what they are talking about.

Flux design is key to success in the fire assay as it is with the mineral works of alchemy in the Dry Ways.

The process of the fire assay begins with the mineral ore finely powdered and dried, from which a sample weighing 29.166 grams is taken. This weight is called an "assay ton"; why it is used will make sense in the end.

To the powdered ore sample, lead oxide (litharge) and a fluxing agent are added and mixed well. This is the tricky part where a little of the artist must step in. The operator must examine the ore and estimate its characteristics such as sulfur content, silica and alumina contents and the overall acidic or alkaline nature of the material. Based on this estimate a fluxing mixture is compounded by adjusting the proportions of a few common salts.

For example, a quartz type ore is somewhat "acidic" and a flux may consist simply of an equal weight of alkali, like sodium carbonate, as there is ore powder. Sometimes a little extra silica in the form of powdered sand is added to keep the blend at about 50 percent silica. In a correct proportion, silica helps lower the melting point of many refractory minerals. To this mixture is added a little flour as a source of carbon which will reduce some of the lead oxide to metallic lead when heated in the furnace (about 1 gram flour will reduce 10 grams of lead). The lead acts as a "collector" of gold and silver from the melt.

So now our well mixed charge to the crucible contains about 30 grams of powdered ore, 30 grams of lead oxide (litharge), 30 grams of sodium carbonate, about 10 grams of silica, and 2 grams of flour.

The crucible is rapidly heated up to about 1100 deg C, during which a series of events will happen. First, tiny particles of lead metal will begin to form throughout the mixture, and the fluxing agents will then start opening the minerals for dissolution in the melt; gold and silver released from the mineral become associated with the particles of lead.

The sodium carbonate melts at about 850 deg C and breaks down silicates, and the litharge melts at about 885 degrees. The particles of lead metal containing now the gold and silver begin to coalesce and fall to the bottom of the crucible like a fine rain, collecting still more of the gold and silver released into the melt.

Once full melt is attained (1100 deg C) the contents of the crucible are poured into a cone-shaped iron mold and allowed to cool. A typical melt may have a top layer of frothy scoria containing mostly the alkali sulfates, then a glassy slag as the main component. The color of this slag tells a lot about what is in the mineral ore other than gold and silver. At the bottom, there will be a lump of lead metal which contains all of the gold and silver from the ore.

This lump of lead is separated easily from the glassy slag by a hammer blow, then pounded into a small cube.

The cube of lead is now placed onto a cupel (pronounced "que – pell"), which is a small solid cylinder having a shallow depression on top to hold the lead. The cupel is usually made of bone ash with a little cement to hold it together. The cupel holding the lead is now placed into a furnace at about 1000 deg C with plenty of ventilation.

In this oxidizing environment, the lead begins to rapidly oxidize into lead oxide again, and what doesn't vapor away, melts and is absorbed by the bone ash. This process of slowly "cooking" off the lead is called cupellation. Some have called this the Bath of Saturn, washing the Sun and Moon; for in the end, when all of the lead has been driven out, there remains a gleaming bead of gold and silver on the cupel.

The weight of this bead, in milligrams, is equivalent to the total weight of gold and silver in the ore expressed as ounces per ton. This is because we started with an "assay ton" of ore powder (29.166 grams), so the calculation is simple and direct.

If we want to know the total of gold and the total of silver separately, we need to separate the two by a process called "parting". Record the weight of the bead, then pound it flat and treat it with concentrated nitric acid. This will dissolve the silver. The insoluble residue is the gold, which is washed with water, dried and weighed. The weight in milligrams will tell us the weight of gold as ounces per ton, and the difference from the total bead weight will tell us the weight of silver.

Illustration from Basil Valentine's *Twelve Keys*. The noble metals, gold and silver, or the king and queen, stand regally as the process of fire assay unfolds before them. In the left lower corner, the fusion is taking place in a crucible. The wolf represents antimony, which we will discuss next. At the lower right corner, Saturn (lead) is poised over a cupel surrounded by flame. In the cupel rests the final bead of the fire assay which contains the gold and silver.

Gold and Antimony

There is another very ancient method of refining gold which you will often come across in alchemical texts and involves the use of antimony, "The Child of Saturn".

One of the interesting properties of antimony is that it dissolves gold quite rapidly. This property was used for purifying gold from other metals used in its alloys.

The impure metal was melted with antimony sulfide (stibnite). The impurities such as copper and silver formed sulfides which could be skimmed off, but the gold went into solution with the antimony as it reduced to metallic form.

The antimony-gold alloy was then heated in a stream of air, causing antimony to volatilize as the trioxide, leaving behind the purified gold much like the cupellation with lead in the fire assay.

The process, often illustrated as a "gray wolf devouring the king", was repeated several times to completely purge the gold of foreign metals as well as to imbibe some of the life force of the antimony into the gold. Due to this property, antimony was sometimes referred to as "Aries of the Philosophers", because the Sun is exalted in the sign of Aries.

> If you would operate by means of our bodies, take a fierce gray wolf, which, though on account of its name it be subject to the sway of warlike Mars, is by birth the offspring of ancient Saturn, and is found in the valleys and mountains of the world, where he roams about savage with hunger. Cast to him the body of the King, and when he has devoured it, burn him entirely to ashes in a great fire. By this process the King will be liberated; and when it has been performed thrice the Lion has overcome the wolf, and will find

nothing more to devour in him. Thus our Body has been rendered fit for the first stage of our work.

Know that this is the only right and legitimate way of purifying our substance: for the Lion purifies himself with the blood of the wolf, and the tincture of its blood agrees most wonderfully with the tincture of the Lion, seeing that the two liquids are closely akin to each other. When the Lion's hunger is appeased, his spirit becomes more powerful than before, and his eyes glitter like the Sun. His internal essence is now of inestimable value for the removing of all defects, and the healing of all diseases. He is pursued by the ten lepers, who desire to drink his blood; and all that are tormented with any kind of sickness are refreshed with this blood.

For whoever drinks of this golden fountain, experiences a renovation of his whole nature, a vanishing of all unhealthy matter, a fresh supply of blood, a strengthening of the heart and of all the vitals, and a permanent bracing of every limb. For it opens all the pores, and through them bears away all that prevents the perfect health of the body, but allows all that is beneficial to remain therein unmolested.

The Twelve Keys of Basil Valentine, Key 1

From Atalanta Fugiens. The king (gold) is devoured by the wolf, but later emerges renewed from the fire.

> Therefore, let men know, that Antimony not only purgeth Gold, cleanseth and frees it from every peregrine Matter, and from all other Metals, but also (by a power innate in itself) effects the same in Men and Beasts.
>
> *Triumphal Chariot*

Alchemical Gold

The texts of rasa shastra describe five types of gold. Three are said to be of celestial origin and tied to the activities of the gods. A fourth type is that obtained from the mines and a fifth type, curiously, is that derived by the transmutation of mercury.

In addition, gold is said to be subject to ten defects which prevent its use in medicine. The ten defects are listed below.

1 Whiteness in color
2 Hardness
3 Non-unctuousness
4 Variegated color
5 Association of impurities
6 Appearance of scales when placed over fire
7 Appearance of blackness in cut surfaces

8   Appearance of whiteness in cut surfaces
9   Cracking while pressing
10   Lightness

If gold is not purified prior to its preparation as a medicine, it causes reduction of strength, loss of potency and suppression of digestive power, and it diminishes enthusiasm, complexion and happiness. After proper purification, gold acts like ambrosia and rejuvenates the body and mind.

In the Western tradition, the sages describe three types of gold. One is celestial, one is inherent in all things and one is the metallic element we usually think of when speaking of gold.

> In order that you may desire nothing that belongs to the theory and practice of our Philosophy, I will tell you that, according to Philosophers, there are three kinds of Gold.

> The first one is an Astral Gold, whose center is in the Sun, which with its rays communicates it, at the same time as its light, to all heavenly bodies under it. It is an igneous substance, and a continuous emanation of solar corpuscles which, being in a perpetual flux and reflux, because of the movement of the sun and the stars, fills up the whole universe. Everything is penetrated by this Gold in the immensity of heavens upon earth, and, in its bowels, we breathe continuously this Astral Gold, the solar particles penetrate our bodies and are ceaselessly exhaled from them.

> The second one is an Elemental Gold, that is, the purest and most fixed portion of the elements, and of all substances composed by them; so that every sublunar being of the three genera contains in its center a precious grain of this Elemental Gold.

The third one is the handsome metal, whose
brightness and inalterable perfection give it a
price, which make it to be considered by all men
as the sovereign remedy for all evils and needs of
life, and as the only foundation of greatness and
human power.

The Hermetical Triumph

Calcination of Gold

Most of the metals are relatively easy to calcine into an oxide
by roasting them in an open dish. Gold and, to a certain extent
silver, remain in the metallic state even after a prolonged
roasting at high temperatures. The preparation of a finely
divided gold dust, called gold calx, is an important form of gold
useful for many preparations.

In this section, we will examine several of the most common
ways of preparing a very fine gold calx. The first two methods
begin with the amalgamation of gold with metallic mercury.

In the first method, we begin by forming a gold and mercury
amalgam by grinding the gold with at least four times its
volume of purified mercury.

Now, squeeze out the excess mercury through a piece of
chamois and retrieve the ball of amalgam from inside the
chamois. Wash this amalgam with salt and vinegar, then finally
with water until it is bright and shiny.

Press the ball of amalgam onto a dish and set it in the sun
to harden.

Pulverize the dried amalgam and mix one part with three
parts purified sea salt, then place the mixture into a strong
distillation flask. Distill the mercury out into a receiver partially
filled with water. Save the mercury for reuse.

The gold calx and salt will remain in the distilling flask.
Wash out the salt with rainwater, then collect and dry the gold
calx. Using the mercury you saved aside, and the dried gold
calx, repeat the whole process two more times.

The final washed and dried gold calx will be exceedingly fine
and ready for use in "seeding" animated mercury or in the
preparation of other gold products.

The second method proceeds as the first method up until the distillation. Instead of distilling the mercury out, it is dissolved out by the use of aqua fortis, nitric acid. The mercury goes into solution as mercury nitrate and the gold calx remains undissolved. The process is repeated two more times using fresh mercury, and the final gold calx is washed thoroughly with fresh rainwater.

The dissolved mercury can be recovered as mercury nitrate by evaporating the solution, or it can be precipitated with sodium or potassium carbonate, dried and roasted into the oxide.

Sodium sulfide can be used to precipitate an artificial cinnabar. The carbonate, oxide or cinnabar can be reduced to metallic mercury again by distillation with iron filings. The mercury nitrate is a powerful oxidizer so be careful with it or you might start a fire or have an explosion.

Also, *Do Not* use an ammonia-based material to precipitate the mercury or you may form the very unstable fulminating mercury and definitely have an explosion. This holds true for gold also; never precipitate gold from a solution using an ammonia based material. Fulminating gold is extremely powerful and unstable when dry.

A simple method for preparing gold calx is by repeated calcinations with common salt. Gold powder, leaf, or filings are mixed with salt and then heated to redness in a furnace. Once cool, the mass is ground finely in a mortar and then returned to the crucible for another heating in the furnace. The cycle of heating and grinding is repeated 7 to 20 times. Finally, the salt is washed out of the gold calx with water several times and the collected powder is gently dried.

In this last method we examine, gold is first dissolved in Aqua Regia. This is the combination of 3 parts hydrochloric acid with 1 part nitric acid. Gently evaporate the solution to obtain beautiful golden orange crystals of hydrochloroauric acid. Minimize exposure to strong light during this process, as the crystals are sensitive and will decompose.

Now dissolve the crystals in distilled water, at least 20 times their volume. Slowly add Oil of Tartar per Deliquiem; that is potassium carbonate which has deliquesced by exposure to the moist night air. Keep adding until the solution is just neutral, by which time you will see a heavy precipitate of gold hydroxide fall to the bottom.

Let this settle, then decant the clear liquid out and wash the gold precipitate with fresh water. Decant as much of the water out as you can, then slowly add hydrochloric acid dropwise until the precipitate just dissolves.

Dilute the solution with water and repeat the precipitation with oil of tartar. Repeat this process of "Solve et Coagula" at least seven times and the final precipitate will be an extremely fine gold calx suitable for preparations such as potable gold.

Potable Gold

Potable gold is literally drinkable gold. This was a much sought after rasayana in the Western world, said to cure most diseases and extend life.

There are many recipes for its preparation, some good, some not so good. The good ones are either fine dispersions of colloidal gold, or the essence of gold; its Sulfur and Mercury, dissolved in a suitable menstruum. The bad ones are just solutions of gold salts which can precipitate out in the body and cause problems.

The following are a selection of some of the more reliable processes you will come across, as described in a number of alchemical works.

The first process comes from Johann Agricolla, in his *Treatise on Gold* written about 1620. This is one of several methods he describes for the extraction of the alchemical Sulfur and Mercury from the body of gold. The final product is a deep golden red tincture containing the essence of gold with powerful healing virtues.

Sulfur and Mercury of Gold

Take some of the best purified gold, as much as you like, have a goldsmith laminate it very thin, the thinner the better. Cut it as big as a Thaler (old coin like the Dollar), the cut round pieces from a stag's antlers, as big and thick as half a

Thaler, take a cement can no wider than the pieces of antlers or half a Thaler, so that only the pieces fit in. One can also make it of good clay, as one pleases. At the bottom of the can put one finger's width of sand or gypsum, which is better. On it put a piece of antler, upon that a piece of your gold, above it again a piece of antler, then again gold. Put everything layer upon layer, as the chymists say, till the can is full or till you have used all your gold. Again, put gypsum upon it till the can is quite full, close the can with good lute, let it dry, then set it in a medium strong cementing fire, at first very gentle, then finally so strong that the can will well glow for one hour or four. Let it cool, open the can, and you will find the gold calcined almost flesh-colored.

This work must be repeated three times. The gold will become quite soft and can be pounded and rubbed. Now mix it with calcined antlers and reverberate it on a cupel but not too strongly, for a whole day. The gold will turn almost the color of bricks. Then it is correctly and well calcined, and you may be sure that you cannot get a better calcination. It will become so subtle that it can easily be used in medicines for several sicknesses without any further preparation, for this calx is sweet and not contaminated by any corrosives.

Upon this beautiful pure calx pour the following prepared menstruum. It will extract its tincture in a few hours like blood, leaving its metallic slime behind. Pour the menstruum off, pour fresh one upon it till all of the tincture is extracted and nothing but a dead earth is left. Nor is that to be thrown out, because it has a special power for drying and cleansing all discharging damages, so that they heal all the sooner. Distill your menstruum down to dryness through sand, and a

244

purple-colored tincture will be left in the glass. Upon that pour a good spirit of wine. How that is to be correctly prepared will be found further on in the Treatise on Tartar. Better, use some quintessence of salt. How to make that will also be taught under its title. Set it closed to digest and it will extract a yet purer tincture. Distill the spirit of wine to half, and you will have a wonderful potable gold. Or, if you pour some quintessence of salt over it, you can leave it such as, without distilling it and use it as a medicine, because the essence of salt is by itself a fine medicine, also without gold, as will also be shown in its proper place.

After the calcination of the gold, I thought of a special menstruum. Now I will also show you how it is to be prepared to make the work and the process perfect.

It depends on the best manipulation, and this is what is to be done: Take a good amount of boy's urine, distill it to half, pour away what is left, and put the distillate again in a retort. Again distill it to half, and do this work three times. With the subtle spirit a beautiful, transparent, shining salt will rise. Rinse all the salt with the spirit out of the alembic, weigh this spirit, mix it with the same amount of the best spirit of wine, let it gently putrefy together for 8 days, then distill it, and you will have a wonderful menstruum for all metals, minerals, and precious stones. With this you can obtain the true tincture of gold.

Agricolla, Treatise on Gold, ch. 2

Oil of Gold from Collectanea Chemica

Jean Battiste Van Helmont, a student of the works of Paracelsus, describes a method for preparing the oil of gold in

Collectanea Chemica, published in 1684. His method is an adaptation of the acetate path.

The preparation employs the calcination of gold using metallic mercury as described above. One part of gold is mixed with five or six parts of purified mercury, then the mercury is distilled out.

The gold calx that remains is again mixed with the same mercury and distilled again. This process is repeated until you find it difficult to get the gold to mix with the mercury.

Take the gold calx and grind it with dry salt until it is very fine. Wash the salt out with rainwater and dry the gold powder. Now the gold is heated to near red hot, then quenched by pouring it into distilled vinegar. Repeat this heating and quenching five or six times, then recover the gold and dry it.

Mix the dried gold with the mercury again and distill the mercury out. Repeat this several times, then recover the gold and repeat the process of heating and quenching in the same distilled vinegar several times.

Now take the vinegar, "which is impregnated with the whole essence of the gold", and filter it to remove all of the gold particles. Very gently evaporate the vinegar and you will obtain a yellow salt.

Dissolve this salt in rainwater, filter, and again gently evaporate to dryness. Place the dried salt into a distillation train, like that used for the acetate distillations, but suitably sized for the small amount of salt.

Proceed with the distillation as described for the acetate work. The white vapor and a "saffron" colored oil will come over into the first receiver and resolve itself into a red liquor. Let the apparatus cool, then remove the liquid and preserve it carefully for use.

> This is one of the greatest medicines under the Sun, and you can hardly get a better. Three drops are able to extinguish any sickness, and in this oyl of gold is the greatest secret of Nature.

This next preparation uses the Philosophical Wine Spirit derived from the acetate path. It is a straightforward and reliable method for obtaining the essence of gold and is one of the methods advocated by Frater Albertus.

The following description is from *Secreta Alchemiae*, written by Kalid ben Jazichi. The text uses A.R. to denote Aqua Regia and S.V. to denote Spiritus Vini. This spiritus vini is the philosophical wine spirit derived from the distillation of lead acetate. Common spirit of wine (alcohol) will not work for this. Commercial acetone does not work very well either, but the wine spirit prepared from distillation of other acetates will work. The distillate from zinc acetate seems to work especially well.

The basic process is to form gold chloride and purify it by recrystallization as detailed above. The resulting crystals are extracted with the "White Wine Spirit" to produce a blood red oil of gold. The White Wine is gently distilled off and the resulting oil purified by dissolving into common wine spirit for use.

Oil of Gold using Philosophical Mercury of Lead

> XXX. An Appendix teaching how to make Aurum Potabile. Take Sal Armoniac, Sal nitre, ana 1 pound: beat them together, and make thereof an A.R.: Then take of the most fine Sol q.v. in thin leaves, and cut into very small pieces, which roll into very thin rolls, and put them into an Urinal, or like Glass, to which put the A.R., so much as to overtop it the depth of an inch.

> XXXI. Then nip up the Glass, and put it to putrefy in Sand, with a gentle heat, like that of the Sun for 3 or 4 days, in which time it will come to dissolution; then break the Glass off at the Neck, and pouring off the A.R. easily and leisurely, leave the dissolved Sol in the bottom, and repeat this work with fresh A.R., 3 or 4 times, and keep the first water, then put on a Helm with Lute, and distill off in Sand: Being cold break the Glass, and

take the Sol, and wash it 3 or 4 times in pure warm water.

XXXII. When the Sol is clean from the A.R., take of it, and put it into the like Glasses, with rectified S.V. 2 or 3 inches above it; put it into putrefaction as before in Sand, stopping the mouth thereof very close for 3 or 4 days; then put the S.V. out, which will be all blood red. If any thing remains in the Glass undissolved, put in more S.V. and let it stand as before. Do this as long as you find any Tincture therein. This is Aurum Potabile.

Secreta Alchymiae, Kalid ben Jazichi

Colloidal Gold and Bhasmas

The physical and physiological properties of materials can change dramatically when they are reduced to a finely dispersed state. There are many cases in which metallic medicines are prepared in this finely divided form called a colloid.

Colloids are dispersions of insoluble particles too small for the eye to see and too light for them to settle out even after prolonged standing. Milk is an example of a colloidal suspension. Much of our biology depends on colloidal materials and they are acceptable to our systems without further processing.

Two of the most popular metallic medicines of this type are colloidal gold and colloidal silver. Today they are generally prepared by electrolytic methods from the pure metals. Their effectiveness is dependent on their particle size; the finer the better.

We generally imagine a solution of a material as a mixture of single molecules dispersed evenly through the dissolving liquid. In reality, the solution consists of clusters of the atoms or molecules of the substance dispersed through the solution. The clusters may be several hundreds of the atoms large. As the clusters are broken down by the processes of "solve et coagula", they become more active and acceptable to the body. When the clusters are broken down until there are single atoms dispersed in the solution, a dramatic change is said to happen. In this monoatomic state, the electronic properties of the atom are said to reconfigure themselves into a stable, high spin state with superconductive properties.

Although the medicinal use of colloidal and monoatomic metals is found in western alchemical tradition, this form of metallic medicine is especially important in the bhasmas of rasa shastra.

Bhasma

This term is from Ayurvedic practice and denotes a mineral or metallic body reduced to extreme subtlety. Literally, *bhasma* means "ash" and the materials are prepared by a complex process of calcination similar to the one described by Agricolla above. Their medicinal virtues have been esteemed for many centuries and even to the present day.

Although the materials required to produce a bhasma are common or easily obtained, the process itself is laborious. There are two main stages each material must pass through and each material has its own needs.

The first stage is the purification of the metal or mineral called "Shodhana". This begins the process of opening the matter and removing characteristic defects and impurities so that its therapeutic virtues come out, making it fit for the next stage.

The second stage, called "Marana", is the actual preparation of the bhasma by a series of calcinations under prescribed conditions rendering them fit for medicinal use.

Svarna Bhasma or gold bhasma is most commonly prepared using arsenic or mercury during the initial phase in order to disperse the metal into fine particles. These two volatile metals (Hg and As) are also used in the preparation of many other metals as well. When done correctly, none of the arsenic or mercury are present in the final product. They pass out as a vapor in the subsequent firings; this is something you will want to avoid exposure to.

The purification (shodhana) of gold is performed by beating the gold into leaves, then heating them over a flame until red hot. The hot gold leaves are then quenched by immersion in a liquid seven times. This is repeated for each of several liquids including sesame oil, buttermilk, cow's milk, cow's urine, and a type of vinegar made from grain. Some recommend a number of herbal decoctions as well, especially Kulattha *(Dolichos biflora)*, a member of the legumes.

This process of heating and quenching in these liquids is held to be a suitable purification for most of the metals in general, though some metals require an additional metal-specific purification method.

This purified gold is mixed with an equal amount of arsenic (or sometimes mercury is used), then triturated with a mortar and pestle for seven days using the juice or a strong decoction of plants to keep a thin pasty consistency. Yes, I did say seven days, and we're just getting warmed up. Some ayurvedic methods require in excess of a thousand hours of grinding to achieve the desired subtlety.

The plants used in the grinding are *Bauhinia variegata* and *Ocimum sanctum* (Linn), common in India. As an alternative, Aloe vera gel and basil can be used instead.

The ground pasty material is now formed into small cakes about an inch in diameter and left in the sun to dry.

Once dry, the cakes are placed between two earthenware bowls, which are then sealed by wrapping with clay-smeared cloth. The whole package is allowed to dry thoroughly, then it is ready for the fire.

In ayurvedic practice, each of the bhasmas is prepared in a firing pit of a specified size and specified fuel to obtain the correct temperature for the required time.

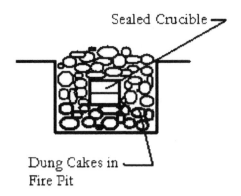

Sealed Crucible

Dung Cakes in Fire Pit

For gold bhasma, a pit which is a twenty-two-inch cube is dug into the ground. The pit is half filled with dried cow dung cakes and the clay-wrapped bowls containing our matter are placed on top, then covered with additional dung cakes to fill the pit. The fire pit is now ignited and allowed to burn completely and finally cool. Retrieve the bowls, break the seal and carefully remove the contents into a mortar.

Now add a quarter of the quantity of arsenic, based on the original weight of gold, and again triturate it using the plant juices as before, for seven days. The resulting pasty matter is formed into cakes and fired as before.

The cycle of firing and grinding is repeated at least seven times and often many more times to achieve a good bhasma. The final three or four firings and grindings omit the arsenic or mercury so these are completely removed by the fire in the end.

The final product will have a pinkish color and be exceedingly fine (compare this with Agricolla's description). If a small amount is sprinkled onto water, it will not sink but instead spread out like oil on water. If a small amount is rubbed between the fingers, the finest lines of your fingerprint will be filled with powder, yet you will feel no grittyness but instead a soft oilyness. As a final test, the bhasma is tested by fire using the usual reducing agents for gold. Success is indicated by the refusal of gold to reduce into metal, which is the state it usually prefers.

Medicinal use of gold bhasma includes cardiac tonic, promoter of eyesight and intellect, and rejuvenator. Counteracting many toxins, and promoting the skin's complexion, it promotes overall longevity and memory.

Gold bhasma has a long history as a curative agent in a wide range of illnesses which plague mankind, including heart disease, tuberculosis, nervous disorders, epilepsy, asthma, anemia and cancer.

The recommended dose ranges from 10 to 25 mg. Twice daily mixed in with butter, cream, milk, or ghee (which is clarified butter; its preparation is described below).

Other bhasmas

The process of making bhasmas forms a key element in ayurvedic Rasa Shastra and provides some of its most powerful alchemical preparations. With this in mind, it may be useful to present a couple more examples of their production. This is a bit of a divergence from our topic on gold, but I think it will supply you with an additional technique for work in the metallic realm which is extremely valuable.

Bhasma of Zinc

Metallic zinc is another important medicine used in ayurveda. You can obtain a relatively pure form of the metal from the casings of old dry cell batteries.

Just strip off the outer labeling and you will find a shiny metallic can which is the zinc. Cut this can down the side and peel it off from the contents. This is a little messy, so work over a large sheet of newspaper and wear gloves.

The material in the can is mainly manganese dioxide with some alkalis, and there is also a carbon rod in the center. Flatten the zinc out and wash it with water and a brush until it is clean and shiny on both sides. Set it aside for now to dry; this is our raw material.

There are a number of different ways to process the zinc into a bhasma; we are going to explore two common methods which are applicable to several other metals as well, most notably the soft and easily melted metals lead and tin.

The first method is a little more difficult but perhaps the most common in use. First we must prepare an agent called "Kajjali", a type of artificial cinnabar, which is widely used in Rasa Shastra for creating metallic medicines. Kajjalli has therapeutic value on its own.

To prepare kajalli, we must first purify a quantity of mercury metal. This is called the "Shodhan of Mercury". Place the mercury into a mortar of porcelain or iron and triturate it with fresh garlic cloves to form a thick paste. This takes about 60 hours of grinding to correctly form the paste, as all traces of the metal must disappear, resulting in a homogenous black paste.

252

Now mix the paste with warm water and the purified mercury will settle out at the bottom of the vessel. Dry the mercury with a paper towel and set it aside.

Next we must perform the "Shodhan of Sulfur". As mercury is known as the "Semen of Shiva", sulfur is known as "The Ovum of Shakti" (Shakti being Shiva's divine consort). To begin, we need to make a quantity of "Ghee", which is clarified butter.

Place about a pound of unsalted, organic butter into a heavy saucepan and let it melt over a medium heat. Soon the butter will begin to foam and crackle as the water boils out and the milk solids separate. Watch it carefully so as not to burn it. After about 20 minutes the water will be gone and the butter will stop crackling and popping, it will become quiet and you will see that it is becoming a beautiful clear golden color as the milk solids settle and turn a tan or light brown color at the bottom. Carefully decant the clear liquid into a container and save it aside as prepared ghee.

Ghee itself possesses remarkable healing properties on its own, both internally and externally; many consider it to be an effective rasayana in doses of less than two tablespoons per day. It also does not need refrigeration and is said to increase in potency with age. Ghee is a favorite cooking aid for sauteing spices in order to release their healing qualities prior to adding them to a main dish. Ghee shares many of the same qualities as Ojas and thus its use increases Ojas in the body.

The next thing we need is a metal pot half filled with milk and covered at the top with a layer of thin cotton cloth secured with a wire band.

Now we are ready to begin the shodhan of sulfur. Place some of the ghee into an iron pot and begin heating (careful not to burn it), then slowly add native sulfur powder while stirring.

Be sure to have adequate ventilation, because the sulfur fumes will get quite strong. Keep adding sulfur until you have a thin pasty mixture that is still pourable and let it cook for a short time. A large part of the sulfur will actually dissolve into the ghee.

Quickly pour the entire content of the mixture through the cotton cloth into the milk. When cool, skim the ghee from the top of the milk and then filter the milk to retrieve the purified sulfur.

If we repeat this process two more times with fresh ghee and milk, the sulfur itself will become a powerful medicine for skin diseases, but one time will be sufficient for our present purpose of making kajalli. The ghee is used to pacify the strong pitta dosha present in the sulfur.

Finally, we take equal parts of the purified mercury and purified sulfur and triturate them together for about 16 hours until we obtain a black powder which is called Kajalli. This is the main agent we will use in preparing the bhasma of zinc.

We are now ready to prepare the bhasma of zinc, which is also called "Jasad Bhasma".

To begin, we must perform the shodhan or purification of our zinc. Heating the strips of zinc and then immersing them in the various liquids as we did for gold is a good general purification. For the metal-specific purification, our clean zinc metal is melted in an iron vessel and then poured into milk or buttermilk (which contains more organic acids). Filter the milk to retrieve the zinc and repeat this process 21 times. I know this all sounds like a lot of work, but nobody said these preparations were easy, just powerfully effective and time tested over several thousand years.

At the end of the 21 pourings, we will have a gray powder which is easily ground and represents the purified zinc. Again, the purpose of shodhan is to begin balancing the doshas of the metal and getting it ready to become "humanized", or fit for human consumption.

The purified zinc is now ready for the second stage of the work, called "Marana of Zinc", which is the actual formation of the bhasma.

Mix equal parts of the purified zinc with kajalli powder and triturate it to a fine, homogeneous powder.

Add fresh lemon juice and triturate for 8 hours. Add fresh Aloe Vera gel and triturate for 8 hours into a thick paste. Mold the paste into small cakes about an inch in diameter and 1/8 to 1/4 inch thick, then let them dry in the sun.

Place the dried cakes into an earthenware dish and cover it with a second dish inverted, then seal the whole thing with clay smeared cloth strips and allow it to dry completely.

Traditionally the composition of the dishes used is important as it provides an additional mechanism whereby impurities are removed from the matter. A typical composition contains 3 parts clay, 1 part of plant fibers from the herb *Crotalaria juncea*, 1 part horse dung, 1 part ash of paddy husk, and half a part of iron rust. These materials are ground, then formed into the various dishes or crucibles and cured in the fire before use. For this experiment we can use CorningWare casserole dishes to good advantage. They are strong and will hold up through many firings without cracking.

Our dried "package" is placed into a fire pit just as we did for gold bhasma and fired using cow dung cakes at a temperature of 800 to 1100 deg C. Of course not everyone is in a housing zone that allows flaming pits of dung in the backyard, so you can utilize a gas or electric furnace or plan a trip to the countryside for the firing.

Once cooled, the contents of the dishes are carefully removed and the process of triturating with lemon juice, then aloe, etc., is repeated for another cycle of firing (note that kajalli is not used again).

Four such cycles is the minimum, and often several more will be required to obtain an acceptable bhasma; some recommend 21 cycles. In addition, note that the mercury used in the initial preparation stage will be long since vaporized and hence not in the final product.

The ground, finished product should pass all of the tests for a bhasma mentioned in the section on gold bhasma. In addition, a specific test of zinc bhasma requires that a small portion be treated with lemon juice whereupon no bubbling should occur.

Zinc bhasma is held to be a powerful medicine for all complaints associated with diabetes, anaemia, asthma, obstinate skin problems, and ailments of the eyes and it helps build muscle tissue. It alleviates pitta and kapha doshas.

Another method used to prepare zinc bhasma is a bit more easily performed as follows.

Begin with the shodhan of zinc as above, heating and quenching in the various liquids, followed by melting and pouring into milk 8 to 21 times.

Melt the purified zinc in an iron vessel, then sprinkle in a pinch or two of powdered turmeric or neem leaf, stirring all the time.

Allow this first addition to burn up completely, then add more of the herb powder and allow it to burn up completely also. Keep stirring and adding herb powder in this manner until the metal has become a yellow powder. Now continue the calcination for an additional four hours until the yellow powder becomes red.

Triturate the red powder for 8 hours with Aloe vera gel and then for 8 hours with lemon juice to form a smooth paste. Form the paste into small cakes and let them dry in the sun. Place the dried cakes into an earthenware bowl; cover it with another bowl and seal with mud-smeared cloth. Once dried, the matter is fired in a pit like we did for gold bhasma.

After cooling, the matter is triturated again with Aloe vera gel, then lemon juice as above repeating the formation of small cakes and firing them. This cycle is repeated at least 8 times, but it may require a full 21 times to form the bhasma which passes all the tests.

Although this form of zinc bhasma is said to be less powerful that that made using mercury, it is a straightforward method of working with the lower melting metals like lead, tin and zinc which avoids the use of mercury and its attendant dangers.

Calcium Bhasma

The last bhasma preparation we will look at is for a mineral product, in this case calcium. Calcium is available to us in a variety of forms. Those derived from the shells of sealife, like oysters, conch, cowrie, pearls and cuttlefish bone as well as the shells of various bird eggs are already a long way towards being "humanized" for medicinal use.

Each of these sources has its particular action on the doshas, for example pearl is said to balance all three doshas and is a rasayana. Coral works on the digestive fire, common sea shells

256

THE BOOK OF GOLD

relieve fevers, while cuttlefish bone pacifies pitta and kapha with a beneficial effect on the eyes.

The preparation of calcium bhasma is relatively easy, so it is a good choice to start with in learning the methodology of these materials.

The Shodhan or purification begins by placing the shells into a container and covering them with a solution of one part fresh lemon juice and four parts rainwater. Let this stand undisturbed overnight. The shells will get a white coating on them. Pour off and discard the liquid, then let the shells dry.

These are now the purified shells ready for the Marana or bhasma preparation. Place the shells into a crucible and seal a lid on it with strips of clay-smeared cloth.

Once dry, the crucible is fired in a pit as described above. They can also be fired in a gas or electric furnace at 800 deg C, up to a maximum of 1100 deg C. Once cooled, the shells are triturated with Aloe Vera gel for eight hours, then spread out to dry. The dried material is now triturated with fresh lemon juice to form a paste. Form the paste into small flat cakes and dry them in the sun. Place the dried cakes into the crucible again and seal as before. Repeat the process of firing, and subsequent triturations at least two more times.

The final bhasma should pass the test of floating on water. In addition, there should be no "itching" on the tongue when it is tasted. Repeat the process until it meets these tests.

During the process of bhasma formation, the calcium carbonate, which exists in the raw material in the form called Aragonite (orthorhombic), passes through a series of intermediate compounds such as calcium oxide and calcium hydroxide. This is the fermentation and putrefaction stage. The final bhasma is calcium carbonate, reborn in the calcite crystal structure (rhombic).

Many of the minerals undergo similar rearrangements of their crystal structure during bhasma production.

Appendix I Chart 9 presents a selection of the most common bhasmas and their associated properties.

Seeds of Gold, the Ferment

> Why minerals alone should be excluded from
> God's primal benediction, when He bade all
> things increase and multiply after their kind, I am
> unable to see; and if minerals have seed they have
> it for the purpose of generic propagation.
>
> Sendivogius, *12th Treatise*

> Our Arcanum is gold exalted to the highest
> degree of perfection to which the combined
> action of Nature and Art can develop it. In gold,
> Nature has reached the term of her efforts; but
> the seed of gold is something more perfect still,
> and in cultivating it we must, therefore, call in the
> aid of Art. The seed of metals is hidden out of
> sight still more completely than that of animals;
> nevertheless, it is within the compass of our Art
> to extract it.
>
> Philalethes, *Metamorphosis of Metals*

There is only one seed of metals. It is "Ojas" of the mineral
kingdom when refined by alchemical art.

> The seed of all metals is the same; but that in
> some it is found nearer to, and in some further
> from the surface, all metallic seed is the seed of
> gold; for gold is the intention of nature in regard
> to all metals. If the base metal is not gold, it is
> only through some accidental hindrance; they are
> all potentially gold.
>
> Philalethes, *Metamorphosis of Metals*

The Alchemical Hermaphrodite

In the alchemical world, everything possesses a Soul, or
Philosophical Sulfur, as one of its Three Essentials, and this
Sulfur has a dual nature, the conscious and the subconscious.

The Ancient Egyptians called this double soul the Ba and
the Ka of a thing. Alchemists have described this by a variety

of opposites or combatants including fighting dragons, the Sun and Moon, The Red Sulfur and The White Sulfur, volatile and fixed Sulfur, and many more.

The Red Sulfur is the solar, male aspect, the growth soul. Responsible for continuity, propagation and multiplication.

The White Sulfur is the lunar, female aspect, the corporeal soul. Nurturing and imparting form and specific character.

The ratio of Red and White Sulfur in a body varies case to case and kingdom to kingdom. In many ways, we can relate these to the "oily" qualities pitta and kapha and their subtle counterparts, Agni and Soma.

Many plants have a powerful Red Sulfur or male aspect. They burst out all over with fresh growth even when repeatedly cut down, but their female soul is weak so they are relatively delicate and soon must hibernate to recoup their energies. A metallic body on the other hand, like lead or copper, has a powerful female soul or White Sulfur. They are heavy and very solidly condensed, but liable to corrosion and unable to repair themselves, much less to grow very quickly at all, because their male soul, the growth soul, is weak.

There is a Dry Way in alchemy designed to unite the best of both kingdoms as an herbal-metallic complex, wherein an herbal soul is transferred to a metallic body as we discussed in the work on metal acetates which is a Wet Way.

The methodology parallels ayurvedic practices of making bhasmas and may in fact form part of the underlying reasoning behind their preparation.

The idea here is that impacts of a soul on matter make it "soul-like". If we subject dead gold with an herb having a powerful male-soul aspect, the herbal soul produces "impacts" on the dead gold until it is revived. The gold is tested by the fire and dies again but is revived with fresh herbal soul.

This cycle of killing then reviving the gold is repeated until the material body is soul-like, subtle and almost waxy in consistency. In a sense, the fiery pitta qualities of the plant are used to awaken and augment the mineral fire much as we use spices to awaken our digestive fire and metabolism.

In practice we take a metal and calcine it with an herb, but the temperature is very critical, because we want to gently drive out the weaker souls of the two materials and retain the stronger souls. These stronger souls will be the male-growth soul of the herb and the female soul aspect of the metal.

This is a true "Chemical Wedding" wherein the opened female soul of the metal accepts the growth inducing soul or male soul of the herb. These two become equal and fuse into unity. This is the "Hermaphrodite Soul". Tejas maturing Ojas into Soma.

> When such a herbo-metallic complex, with an hermaphrodite soul, is taken as a drug, the accepting system never lacks the presence of soul and is therefore bound to become everlasting: thus man can become immortal.

> When the same substance is seeded into mercury, the latter becomes everlasting as metal; which is gold. The resultant gold is also the carrier of an ever-growing soul so that the same transferred to a potful of mercury, in turn, changes the latter into gold.

> Thus Alchemical Gold, like the original herbo metallic substance, is an Hermaphrodite by constitution and a Ferment by function.
>
> Mahdihassan, *Parachemy*

The Golden Chain of Homer describes the process of "sweetening" mineral and metallic substances, which brings them to a state fit for human consumption.

"You can't move from one extreme to another extreme without the proper medium."

In order to pass from one extreme, the mineral realm, to another extreme, the animal realm, we make use of the medium between then, which is the vegetable realm.

The process takes some time to complete. We used the same techniques in preparing bhasmas. Take gold filings or the filings of some other metal you want to work with. It may be wise to practice this with something cheap like steel wool for iron or a ball of thin copper wire for copper.

The metal will form the core for a ball. Wrap the core of metal with fresh chopped herb. The herb should have a strong growth soul. You can also use herbs under the same astrological rulership as the metal. Layer the herb on the metal at least half the metal core diameter thick, then wrap the ball with a piece of cheesecloth to keep it together.

Now place layers of clay smeared cloth strips over the whole surface of the metal/herb ball one quarter to half an inch thick. A low fire clay (cone 4 to 6) works well; you can find it at any pottery supply store. Set the finished ball in a cool place to dry very slowly. If cracks appear in the clay, you will have to seal them by using a wet knife or wooden tool. As an alternative, you can seal the herb/metal ball into two earthenware bowls like we did for the bhasmas.

Once you are sure the ball is completely dry, it's time to calcine the contents. The traditional way is to dig a small pit in the ground, line it with fuel to burn at the bottom, then add your clay ball and cover it completely with more fuel to the top of the pit.

Set the fuel on fire and let it maintain for at least four hours, then let the fire go out and cool. The best fuel for this is said to be horse or cow dung pressed into cakes and dried because it provides the right temperature (about 800 deg C) and quality of heat, whereas charcoal burns too hot. Of course our "Philosophical Egg" can be fired in an oven with an exhaust fan as well.

Retrieve the clay ball from the fire pit; hopefully it didn't crack open. Carefully crack open the ball now and gather together the calcined metal residue from inside. Grind the residue with a mortar and pestle and then you are ready to begin another cycle, where the ground residue becomes the core of the fresh herb and clay ball.

Once the matter gets to the state where it is fully powdered, you can begin to triturate it with a very concentrated decoction of the fresh plant. Form the resulting paste into small round cakes and let them dry in the sun. Now continue the process of firing with these cakes, trituration, and refiring.

It may take up to 40 cycles and more for success in forming the Hermaphrodite Soul. The matter should become very fine and unctuous. The plant salts have driven the metal into a very fine colloidal state.

The Golden Carbuncle of the Ancients

Rudolf Glauber describes another way of uniting vegetable and metallic natures in his work *Centuries*. He mentions "the manner of conjoining gold with any burning and volatile vegetable sulfur" to produce a red extract which is "nearly as good as potable gold".

The method is fairly easy and requires only a small amount of gold, a vegetable charcoal and his "Sal Mirabilis".

Today, Sal Mirabilis is known as sodium sulfate, though it still carries the common name of "Glaubers Salt".

Glauber studied and wrote extensively regarding this salt as a great arcanum for medicine and alchemical uses. Current medicine considers it only as a mild laxative, but Glauber describes it as a key to opening the mineral realm and releasing great healing potentials.

If you want to prepare Sal Mirabilis in the old way, simply treat sea salt with concentrated Oil of Vitriol (sulfuric acid) using a glassware setup as shown below.

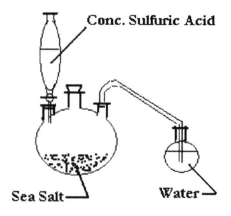

The sulfuric acid is slowly dripped onto sea salt and the reaction produces hydrogen chloride, called "spirit of salt", as a gas which is bubbled into water.

Water will dissolve large volumes of hydrogen chloride, and the resulting solution will be hydrochloric acid, which is useful in many alchemical works so save it aside tightly sealed.

The sulfuric acid treated salt remaining in the flask has now been converted into sodium sulfate or Sal Mirabilis. This salt is now recrystallized from rainwater several times before use. The salt looks like crushed ice and melts at an incredibly low temperature of 32 deg C.

The crystals form in the orthorhombic habit and hold a large amount of water which must be removed for our use here. Dry the crystals at 100 deg C until they become pure white.

In this form the crystals melt at about 800 deg C. By the way, the salt is also insoluble in alcohol and is frequently used to dry alcohol of residual water, similar to the use of salt of tartar.

Now that we have our starting material, we can proceed to make what Glauber calls "The Golden Carbuncle of the Ancients".

For one part of gold, melt eight to ten parts of the sal mirabilis in a crucible (parts by volume). Once it is fused, slowly add the gold as a powder or thin leaves. Slowly add small pieces of charcoal to the melt.

The best charcoal for this is from grape vines, but you can use oak, cedar, juniper, beech, or boxwood also. Continue the fusion for 15 to 30 minutes, then cast the melt out onto a hot metal dish to cool.

The mass will have a deep red color due to the finely divided gold. This is the "Golden Carbuncle" that "shines in the night like a burning coal".

Powder the mass and extract with good spirit of wine. The extract will become deep red, with properties similar to good potable gold. The residue from the extraction can be put through the process again to create more of the red extract which is a form of colloidal gold.

Rasa Shastra and Mercury

The transmutation of lesser metals into gold has never been relegated to the realm of myth and legend in India. Though difficult to attain, there are reports of public demonstrations.

In the Birla Temple of New Delhi stand two marble plaques carved in Sanskrit which give testimony to public demonstrations held in 1941 and 1942 before government dignitaries, prominent industrialists, and scientists. The plaques detail how a master of Rasa Shastra, using just a small amount of powder derived from metallic mercury, which had undergone arduous processing, successfully transmuted nearly a kilogram of common mercury into fine gold.

Metallic mercury was known by a variety of interesting names including "Maker of Gold", "Seed or Semen of Shiva" and "King of Rasas".

The preparation of mercury to be used medicinally and for the transmutation of metals must undergo eighteen "samskaras", or complex treatments.

Common mercury from the mine is considered to have five defects and seven layers of adulterants which must be carefully removed before it is fit for use in medicine or for transmutation.

Briefly, the eighteen processes with mercury as the central element are listed below. If mercury is to be used in the preparation of other metallic medicines and not for transmutation, only the first eight processes are considered essential.

1. Svedana: Steaming mercury over a type of grain vinegar after mixing it with a number of plant substances such as ginger and black pepper along with rock salt for 21 days. This begins to loosen the impurities inherent in the mercury.

2. Mardana: Trituration (grinding) mercury in an iron mortar along with various plants, sugar, salt and acidic materials for 3 to 21 days, then washing. This removes external excretions.

3. Murchana: Triturating mercury in a mortar with more plant extracts till it loses its cohesiveness and breaks into small globules. This removes still more excretia and the mercury begins to lose much of its toxic nature.

4. Uthapana: Steaming mercury again along with alkalis, salts, myrobalans, and alum, then grinding the mercury again in sunlight so that the characteristics of mercury, freed from impurities, are restored.

5. Patana: Grinding mercury with alkalis and salts, then subjecting the product to distillation of three types, ie., upwards, downwards, and sideways. This removes yet more of the imbalances present in mercury.

6. Rodhana: Mixing the distilled mercury with salt water and sal ammoniac, then heating in a closed pot for 21 days. This helps to restore the therapeutic "vigor or potency" of mercury and also makes it more thermostable.

7. Niyamana: Steaming the mercury for three days after making it into a paste with a number of plant products including garlic, tamarind, and salt. The paste is then heated together with sal ammoniac and lemon juice inside a crucible made of rock salt. This helps to restrain the mobility of mercury and increases its luster.

8. Dipana: Steaming this product with alum, black pepper, sour gruel, borax and iron sulfate along with some vegetable substances for 3 days in order to "kindle" the desire of mercury to attain the power of assimilation.

9. Grasa Mana: Beginning of the fixation and assimilation of the ash (bhasma) of biotite mica and of gold as a seed to cultivate the transformative power of the mercury.

10. Carana: Boiling this product with sour gruel, leaves of certain plants, alum and other salts for a week so that the mica is fully assimilated.

11. Garbha Druti: Heating and grinding mercury with a variety of catalytic agents including alum, iron sulfate and sulfur so that the "essence" of the mica and gold

becomes liquified and the resultant, after cooling, passes through a piece of cloth.

12. Bahya Druti: Trituration with plants and goat urine for 100 times in order to enhance the potency of the mercury now containing the "essence" of the added mineral and metallic substances.

13. Jarana: Heating the mercurial product with the desired minerals or metals, alkalis and salts so that they are completely digested and assimilated, making the mercury suitable for curing disease and promoting rejuvenation.

14. Ranjana: A complex process involving the treatment of mercury with sulphur, gold, silver and copper as well as various salts in such a way that mercury attains color.

15. Sarana: Digesting mercury with gold or silver in an oil base to increase its ability to effect transformation.

16. Kramana: Forming a paste of the mercury with several plant extracts, magnetite, cinnabar, calamine, and milk, then heating it carefully with a view to increasing its power of penetration into tissues and metals to effect transmutation.

17. Vedhana: Forming a paste of the resultant mercury with a few select substances including oil and applying it to base metals, then heating in order to effect transmutation. This is the final test of the product.

18. Bhaksana: Consuming a prescribed quantity of the mercurial product which has undergone the foregoing 17 processes, for rejuvenation and longevity.

This sequence for processing mercury was painstakingly followed by Indian alchemists; but there were also variations possible in the choice of plants or their extracts, salts, alkaline and acidic substances, minerals and other added ingredients as well as processing times and temperatures.

The ultimate goal which lay behind all of this was that the mercurial product, after undergoing sequentially the first seventeen processes, was considered to have all the powers of transmutation. At this stage, it was to be tested for its efficacy in transmuting base metals into gold and, if the test was positive, it was to be used for the eighteenth process, which is the transformation of the alchemist himself. The final product, if consumed in proper quantity would, it was claimed, rejuvenate the body in such a way that it would become as "resplendent and imperishable as gold".

Analytical Data

Appendix II D contains a selection of instrumental scans of the oil derived from gold.

APPENDIX I

CHART 1. **Properties of the three Mahagunas**

Sattva	**Rajas**	**Tamas**
Potential energy	Kinetic Energy	Inertia
Equilibrium	Action	Decay
Essence	Mobile	Heaviness
Consciousness	All Movement	Unconsciousness
Understanding	Cognition	Sleep

These are the essential qualities of consciousness. Sattva gives rise to the mind and senses as well as clarity of perception. Rajas is responsible for all forms of movement, and Tamas gives rise to the "Five Elements" and the sensory inputs of sight, smell, taste, touch, and hearing.

CHART 2. **The Qualities of Nature expressed through the Doshas**

"Mercury" VATA	"Sulfur" PITTA	"Salt" KAPHA
Dry	Liquid	Liquid
Cold	Hot	Cold
Light	Light	Heavy
Subtle	Subtle	Dense
Mobile	Mobile	Static
Clear	Clear	Cloudy
Rough	Oily/ Spreading	Smooth/Slimy
	Sharp	Soft or Hard
		Oily
		Sticky
		Gross

CHART 3. **Dosha Types**

Determine Your Constitution

Characteristic	VATA	PITA	KAPHA
Frame	thin, irregular	moderate	thick
Bones	light	medium	heavy
Shoulders	narrow	medium	broad
Chest	thin, small	medium	broad
Weight	low	moderate	heavy
	difficult to gain	hard to gain	gains easily
Skin	dry, cold	oily, warm	oily, cool
Skin tone	brownish	pink	white
Head	narrow	medium	broad
Hair	dry dandruff	oily, thin	slightly oily
	kinky, curly	blonde, straight	thick, wavy
	dull	graying, bald	strong
Teeth	crooked	gums bleed easily	strong
Lips	thin, small, dry	medium, soft, red	full, smooth
Eyes	small, active	sharp, penetrating	large
	busy	intense, burning	attractive
	dry	bloodshot	watery
Eyebrows	thin, dry, scaly	moderate	thick
Hands	small, thin	moderate	large, thick
	dry, rough	pink, moist	oily, firm
	cold	warm	variable

Appetite	variable	excessive	moderate
Hunger every	two hours	four hours	six hours
Reaction time	quick	moderate	slow
Thirst	variable	excessive	scanty
Stool	dry hard	soft	thick, oily
	constipated	loose	slow
	every 1-2 days	2-3 daily	1 a day
Urine	scanty	profuse	moderate
	colorless	yellow, strong	milky
Sweat	rare	profuse	moderate
Menstruation	scanty, painful	moderate to heavy	regular
Activity level	very busy	moderate	little
Endurance	low	moderate	lots
Sleep	restless	little but sound	deep, prolonged
Prefered climate	warm, wet	cold	warm, dry
Mind	restless, active	intense	calm, slow
	lively, curious		
	easily influenced	influential	hard to influence
Emotions	fearful	impatient, angry	indifferent
	anxious	jealous	love, affection
	cool	warm	steady
Beliefs	changeable	fanatical	steady
Memory	recent, good	sharp	slow
	long term poor	selective	prolonged
Dreams	flying	colorful, intense	watery

	running away	passionate	romantic
Speech	quick, talkative	sharp, convincing	slow, definite
Money	poor	moderate	rich
Spends on	everything	quality	acquisitive
Career	inventive	engineer	manager
Sex drive	mental	active	slow to arouse
Remembers	sound	sight	feeling
Reaction	anxiety, fear	anger	denial

CHART 4. **Effects of Taste on the Doshas**

Dosha	Increased by	Decreased by
Vata dry, cold, light	Astringent Bitter Pungent	Sweet Sour Salty
Pitta hot, light, wet	Pungent Sour Salty	Bitter Astringent Sweet
Kapha heavy, wet, cool	Sweet Salty Sour	Bitter Pungent Astringent

CHART 6. **VPK Factors**

Herb	Planet	Taste	Energy	Effect on Doshas
Alfalfa	Saturn	sw	cold	V+ P- K-
Aloe	Mars	bit	cold	VPK=
Angelica	Sun	pun	hot	V- P+ K-
Anise	Merc.	pun	hot	V- P+ K-
Basil	Mars	pun	hot	V- P+ K-
Blk Pepper	Sun	pun	hot	V- P+ K-
Burdock	Venus	bit	cold	V+ P- K-
Calendula	Sun	bit	cold	V+ P- K-
Caraway	Merc	pun	hot	V- P+ K-
Chamomile	Sun	bit	cold	Vo P- K-
Chickweed	Moon	bit	cold	V+ P- K-
Cinnamon	Sun	pun	hot	V- P+ K-
Comfrey	Saturn	sw	cold	V- P- K+
Coriander	Mars	bit	cold	VPK=
Cumin	Mars	pun	cold	VPK=
Dandelion	Jupiter	bit	cold	V+ P- K-
Elder	Venus	bit	cold	Vo P- K-
Eyebright	Sun	bit	cold	V+ P- K-
Fennel	Merc	sw	cold	VPK=
Garlic	Mars	all	hot	V- P+ K-
Ginger	Sun	pun	hot	V- P+ K-
Hawthorn	Mars	so	hot	V- P- Ko
Hops	Mars	bit	cold	V+ P- K-
Horsetail	Saturn	bit	cold	V+ P- K-
Juniper	Sun	pun	hot	V- P+ K-
Lavender	Merc	pun	cold	Vo P- K-
Lemon	Sun	so	cold	V- P- Ko
Licorice	Merc	sw	cold	V- P- K+
Marjoram	Merc	pun	hot	V- P+ K-
Melissa	Jupiter	sw	cold	Vo P- K-
Mullein	Saturn	bit	cold	V+ P- K-
Myrrh	Sun	pun	hot	V- P+ K-
Nettle	Mars	ast	cold	V+ P- K-
Nutmeg	Moon	pun	hot	V- P+ K-

Oregano	Merc	pun	hot	V- P+ K-
Parsley	Merc	pun	hot	V- P+ K-
Plantain	Venus	ast	cold	V+ P- K-
Poppy seed	Moon	pun	hot	V- P+ K-
Rose	Venus	bit	cold	VPK=
Rosemary	Sun	pun	hot	V- P+ K-
Saffron	Sun	pun	cold	VPK=
Sage	Jupiter	pun	hot	V- P+ K-
Senna	Mars	bit	cold	V+ P- K-
Skullcap	Merc	bit	cold	Vo P- K-
St John's wort	Sun	bit	cold	V+ P- K-
Thyme	Venus	pun	hot	V- P+ K-
Turmeric	Moon	pun	hot	V- Po K-
Valerian	Merc	pun	hot	V- P+ K-
Walnut	Sun	sw	hot	V- P+ K+
Watercress	Moon	pun	hot	V- P+ K-
Willow	Moon	bit	cold	V+ P- K-
Wormwood	Mars	bit	cold	Vo P- K-
Yarrow	Venus	bit	cold	V+ P- K-

CHART 7. **Rasayanas for the seven tissues**

Rejuvenatives and Vehicles for the 7 Dhatus

Dhatu	**Rasayana**	**Anupana**
Rasa	Tulsi	Milk Fresh Ginger
Rakta	Manjistha Neem	Pomegranate Juice
Mamsa	Ashwagandha Bala	Ghee and Honey
Meda	Kutki Chitrak	Honey Hot Water
Asthi	Guggulu: Dashamula	Milk
Majja	Brahmi Jatamamsi	Ghee
Shukra (male)	Kapikacchu Vidhari	Milk Draksha
Artava (female)	Ashok	Aloe Vera Gel

CHART 8. **Planetary Seals**

Planetary Seal of the Sun

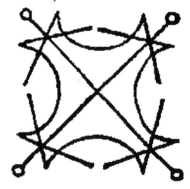

Planetary Seal of the Moon

Planetary Seal of Mars

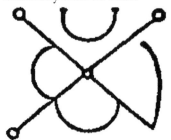

Planetary Seal of Mercury

Planetary Seal of Jupiter

Planetary Seal of Venus

Planetary Seal of Saturn

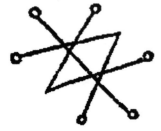

CHART 9. **Table of Common Bhasmas**

Material	Taste	Energy	V	P	K	Dose(mg)
Gold	Sweet	Cold	-	-		30 to 120
Silver	Astr.	Cold	-	-		30 to 60
Copper	Bit.	Hot			-	60 to 120
Iron	Astr	Cold		-	-	125 to 250
Tin	Astr	Hot	-		-	60 to 240
Lead	Bit	Hot	-		-	60 to 120
Zinc	Astr	Cold		-	-	60 to 240
Biotite	Sweet	Cold	-	-	-	15 to 240
Pyrite	Pung.	Hot	-	-		60 to 120
Copper Sulfate	Pung	Hot	-	-	-	15 to 30
Sulfur	Pung	Hot		-		120 to 960
Hemtite	Sweet	Cold		-		240 to 480
Iron Sulfate	Astr	Hot	-		-	60 to 240
Alum	Astr	Hot	-	-	-	240 to 480
Realgar	Bit	Hot	-		-	5 to 8
Stibnite	Sweet	Cold	-	-		30 to 60
Sal Ammoniac	Salty	Hot	-	-	-	60 to 120
Cowrie Shell	Bit	Hot	-		-	250 to 500
Cinnabar	Pung	Hot	-	-	-	60 to 240
Litharge	Astr	Cold	-		-	External
Iron Oxide	Astr	Cold		-	-	30 to 240

Note: A minus sign (-) under V, P, or K indicates a passive or decreasing influence on Vata, Pitta, or Kapha.

CHART 10. **Temperature and Color**

750° F; 480° C faint red
850° F; 580° C dark red
1000° F; 730° C bright red
1200° F; 930° C bright orange
1400° F; 1100° C pale yellow-orange
1600° F; 1300° C yellowish white
above 1700° F or 1400° C white

APPENDIX II

Introduction to the Analytical Section

This appendix provides a very small selection of analytical scans obtained from some of the products described in the alchemical works. They are not meant to be detailed chemical analyses, as that will encompass a work in itself. Nor do they address medical applications or toxicity issues. They are provided to illustrate the nature and complexity of the various materials and lend some guidance to future researchers.

From the alchemical viewpoint, these materials represent vehicles for subtle principles. Just as alcohol is considered the vehicle for the vegetable spirit, these are the vehicles of mineral essence. The alcohol is not the spirit, just a body housing it; a magnet holding an invisible power.

With that in mind, these first few pages present an overview of the technology involved in the analysis and what all of these strange looking graphs tell us.

There are three main types of information presented as analytical scans. These include infrared spectra, gas chromatograms, and ion chromatograms from mass spectrometry. Information on metals content was provided by inductively coupled argon plasma optical emission spectrometry, which happily is abbreviated as ICP.

Infrared spectra provide information on molecular structure. The bending and stretching of chemical bonds between atoms in a compound occur at frequencies in the infrared region of the electromagnetic spectrum.

A compound exposed to infrared energy will absorb those frequencies which resonate with the various chemical bonds present. By examination of the frequency ranges absorbed, we can determine the types of chemical bonding present and then deduce a structure of the compound.

The table below presents a simple guide to the frequency ranges for a selection of common atomic bonds.

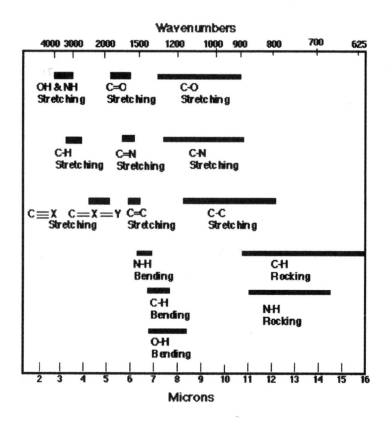

As an example, the infrared spectrum of ethanol is shown below. The broad absorption around 3400 is due to the OH portion of the molecule, stretching back and forth from its attachment to the carbon atom. This broad absorption is characteristic of alcohols and phenols. Its peak of absorption will vary depending on other aspects of the compound's geometry.

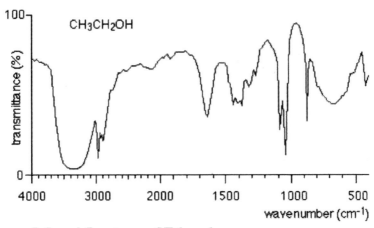

Infrared Spectrum of Ethanol

The second example of an infrared spectrogram is from acetone. The strong absorption around 1700 is characteristic of the double bond stretching between carbon and oxygen (C=O) as found in ketones, esters and organic acids. Again, the fine structure of the molecule will shift the peak absorption around the narrow range of about 1600 to 1900 as shown in the table above.

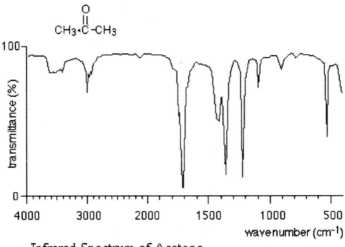

Infrared Spectrum of Acetone

The second and third types of analytical information depend on the principles of chromatography as a method for separating the compounds present in a complex mixture.

You have seen this in everyday life, but probably did not recognize it as such. You spill a drink on your paperwork and soon observe bands of red, blue and green, bleeding out with the liquid. These are the various pigments that were blended together to create the ink. Some of the bands like the paper and do not move very far from the original writing. Other colors like the liquid and move rapidly along with it. Still other colors are torn between liking the paper and liking the liquid, so they move slowly along the paper out from the point of origin.

Gas chromatography uses a similar process to effect the separation of compounds in a mixture. In this case, the paper is replaced by a long piece of tubing filled with a special chemical packing, while an inert gas, like helium, is the motive force pushing the material through the heated column. Some compounds prefer to move rapidly along with the gas, while others like the column packing material to varying degrees. The columns can be quite long, 50 to 100 meter coils, and the various compounds emerge at the other end one by one, fully separated from the others.

As the compounds emerge from the column, they pass into a small flame of hydrogen burning between electrified metal plates. The compound burns and produces charged particles which alter the electrical field between the plates, and subsequently this change is recognized as a signal and plotted on a graph. The relative size of the signal provides information on the amounts of each compound present.

There are many types of column packing material and many variable parameters, such as gas flow and heating conditions, which can be optimized to obtain a good separation of the compounds in a mixture.

Shown below is a simple chromatogram for the essential oil of the herb Melissa. Each of the peaks shown indicates a separate compound present in the oil and its relative concentration. What we recognize as the sweet smell of melissa is the blending of these various compounds.

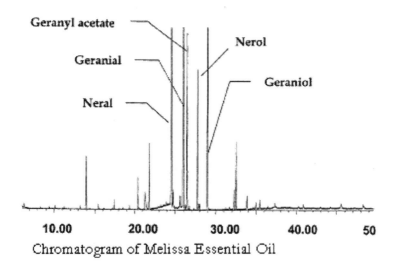

Chromatogram of Melissa Essential Oil

It takes a lot of trial and error, using known compounds, to determine when and where a compound will emerge from a particular chromatographic system.

Instead of passing the emerging compounds through a flame, they can be introduced into a mass spectrometer.

This combination of techniques is called gas chromatography with mass spectrometric detection, abbreviated GC/MS.

As compounds emerge from the chromatographic column, they are subjected to an electrical discharge which breaks the compound into charged fragments. The fragments are accelerated by an electrical field, then passed through a magnetic field which alters their trajectories based on their total mass. Detectors tally up the number of fragments of the different masses and present the result as a total ion chromatogram. Based on this information, a structure of the compound can be deduced.

As an example, the total ion chromatogram of a red wine is shown below, followed by a list of compounds determined by computer matching with a library of known compounds.

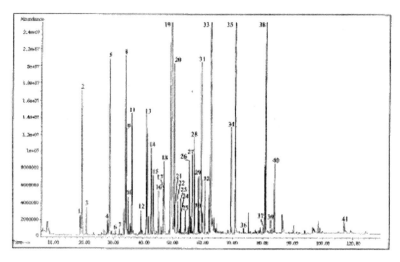

Ion Chromatogram of Red Wine

List of compounds

1	2-Methyl-1-Butanol	18	Ethyl Caprate	34	Diethyl Malonate
2	3-Methyl-1-Butanol	19	Diethyl Succinate	35	Caprylic Acid
3	Ethyl Caproate	20	α-Terpineole	36	Ethyl Cinnamate
4	Ethyl Lactate	21	Terpenediol-I	37	Butyric Acid-i-Butyl Ester
5	Hexanol	22	Methionol	38	Capric Acid
6	3-Hexene-1-ol	23	Citronellol	39	Phenyl Acetic Acid-i-
7	2-Hexene-1-ol	24	cis-p-Linalool Oxide		Butyl Ester
8	Ethyl Caprylate	25	trans-p-Linalool Oxide	40	trans-Geranylic Acid
9	cis-(f)-Linalool Oxide	26	Benzene Acetic Acid Ethyl	41	Myristic Acid
10	Octene-1-ol-3		Ester		
11	trans-(f)-Linalool Oxide	27	Nerol		
12	3-OH-Ethyl Butyrate	28	Phenyl Ethyl Acetate		
13	Linalool	29	Ethyl Laurate		
14	Acetoine	30	Geraniol		
15	i-Butyric Acid	31	Caproic Acid		
16	Hotrienol	32	Benzene Methanol		
17	Butyric Acid	33	2-Phenyl Ethanol		

In a nutshell, this is the technology behind the various scans presented in this appendix.

This is a unique look into the world of practical alchemical work and hopefully will lay the groundwork for future researchers.

APPENDIX II A

The Acetate Path

The following series of analytical scans examines the products of a typical dry distillation on the Acetate Path.

The subject mineral is sodium acetate prepared from a natural source of sodium carbonate called "Natron". This mineral was chosen for its relative non-toxic nature and high yield of oil and spirit.

Shown below is an infrared spectrogram obtained from the purified sodium acetate ready to be distilled.

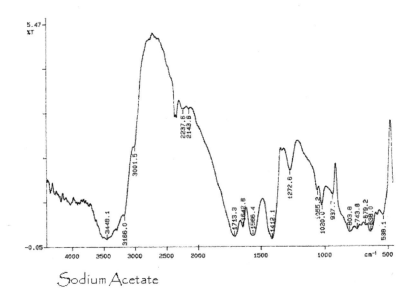

Sodium Acetate

A time VS temperature plot from a distillation of sodium acetate is shown below. The upper line follows the temperature inside the stillpot, while the lower line follows the temperature at the distilling head.

The sodium acetate liquified below 100 degrees and the "phlegm" distilled rapidly. The zig-zag in the stillpot temperature curve at about 65 minutes and 200 degrees is where the sodium acetate suddenly resolidified and began to rapidly heat up. At this point, the white vapor began filling the apparatus and the Menstruum Foetens condensed in the receiver as a red-amber liquid.

The sharp increase of the head temperature near the end is where the blood red oil began coming over.

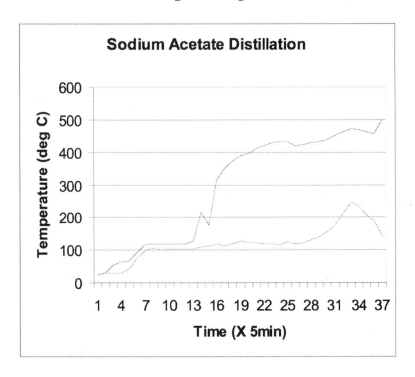

The table below lists the distillation products for sodium acetate. Using a 1-liter distilling flask, an optimum quantity of 400g sodium acetate was found to give fairly consistent results.

Trial	Wt. of Acetate	Vol. Phlegm (ml)	Vol. Menstruum Foetens (ml)	Vol. Cold Trap (ml)	Solid Residue
1	400g	162	50	2	150g
2	300g	124	40	1	110g
3	500g	152	68	3	148g
4	400g	158	65	2	141g
5	400g	148	74	1	140g
6	400g	156	61	2	154g

The collected "Phlegm" and "Menstruum Foetens" were analyzed by gas chromatography and mass spectrometry as presented below. The Phlegm proved to be mainly water, which was associated with the crystalline sodium acetate. In addition to the water, acetic acid was present as 10 to 20 percent of the volume collected. A trace amount of acetone was also detected.

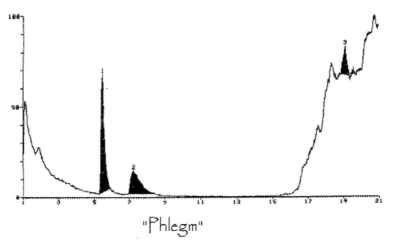

"Phlegm"

The Menstruum Foetens was found to be much more complex, similar to an essential oil from plants.

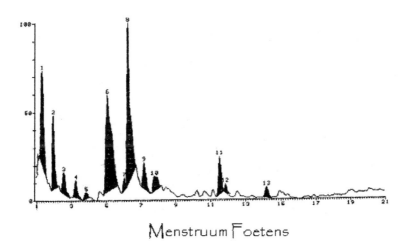

Menstruum Foetens

289

GC-MS analysis indicates the presence of acetone as a major constituent, along with a variety of aromatic compounds such as 1,2,3-Trimethyl Benzene, 2,6-Dimethyl Phenol and 2-Ethyl Phenol.

An infrared scan of the Menstruum Foetens shows its phenolic nature.

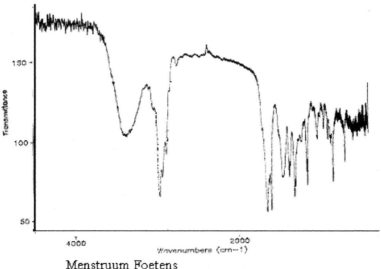

Menstruum Foetens

In an attempt to isolate fractions from the Menstruum Foetens for clear identification, a portion was subjected to slow fractional distillation. The distillates seemed to form their own natural breaks or ranges in which they came over.

First was the separation of the "White Wine" from the "Red Wine". This occurred at 56 deg C, which is the boiling point of acetone, and represented 67% of the initial volume of Menstruum Foetens placed into the flask for distillation.

This was followed by a fraction which distilled between 96 and 100 deg C. On standing, it was discovered that two fractions had codistilled together and were immiscible. A clear fraction settled to the bottom and a golden fraction floated on top. These two fractions were easily separated with a separatory funnel.

The fourth fraction came over between 100 and 130 deg C and was quite acidic in nature. The fifth fraction collected came over between 210 and 219 deg C with a strong aromatic scent.

Each of the collected fractions were examined by gas chromatography, optimized for maximum separation, as shown below.

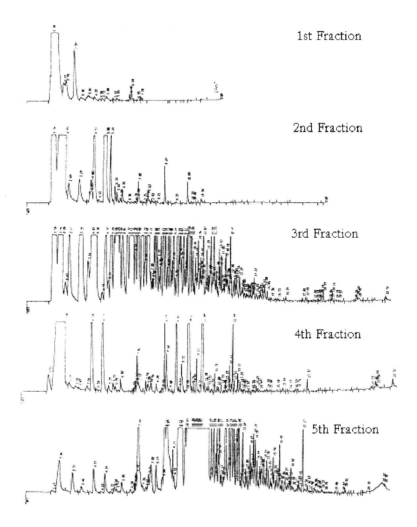

1st Fraction

2nd Fraction

3rd Fraction

4th Fraction

5th Fraction

One can readily see the complex nature of the Menstruum Foetens, even after fractional distillation into five parts.

The first fraction represents Becker's "Spiritus Aceti Oleosus", containing not only acetone, but some of the volatile oils as well. This is the "White Wine Spirit" so useful in extracting tinctures from other minerals.

Each of the collected fractions was examined by infrared spectroscopy as shown in the following scans.

The first fraction confirms the presence of acetone accompanied with traces of phenolic substances.

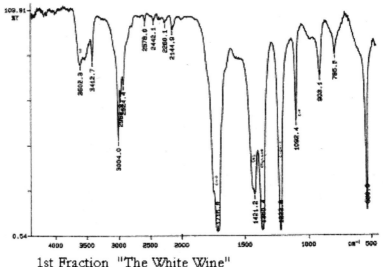

1st Fraction "The White Wine"

Fraction two, which was the clear distillate that codistilled with the golden fraction (Fraction 3), was very acidic. It represented 7% of the original liquid distilled. This proved to be more than just acetic acid. Titration with a standard alkali indicated the possibility of a tribasic acid whose identity remains to be verified. It definitely attacks and dissolves many of the common metals.

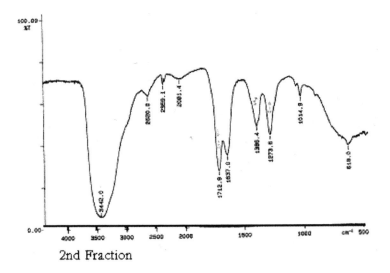

2nd Fraction

The third fraction seemed to be unique in the way it separated itself from the second fraction by collecting itself to the top layer as a golden oily liquid. This fraction was 5% of the Menstruum Foetens' original volume.

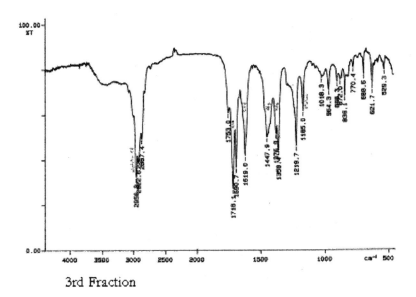

3rd Fraction

Initial examination of the liquid indicated the presence of a neutral polar compound, possibly an alcohol, ketone, or aldehyde.

This third fraction was subjected to fractional distillation and seemed to split itself into four convenient temperature ranges.

In the hope that these were now four easily identifiable compounds, they were subjected to gas chromatography in order to confirm their relative purity.

As one can see, from the resulting chromatograms of Fraction 3 A through D, these were anything but simple compounds. It seemed the closer one looked, the more complex things became.

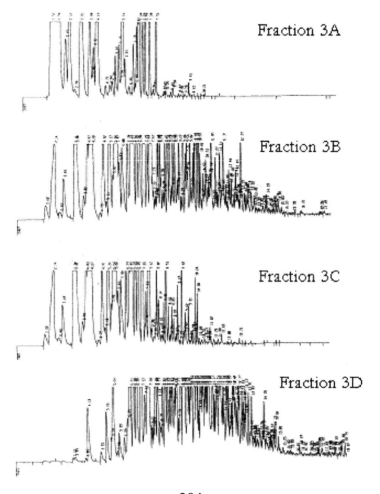

Fraction 3A

Fraction 3B

Fraction 3C

Fraction 3D

Derivatives that were isolated from this third fraction indicate the presence of Isophorone, Mesityl Oxide and 2-Methyl Propanone. Obviously, a great many more compounds remain to be determined.

The fourth fraction obtained from the Menstruum Foetens distilled over between 100 to 130 deg C and represents 12.5% of the original volume. It was golden yellow in color with a very sharp odor. Similar to the second fraction, the major constituent was determined to be acetic acid.

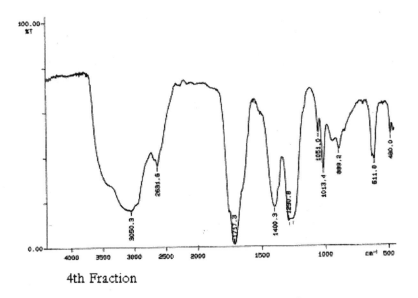

4th Fraction

The fifth fraction obtained was a viscous golden oil having a strong aromatic odor. This represented 5% of the original volume of Menstruum Foetens.

The infrared scan shown below indicates a strong presence of phenolic substances. This was confirmed by isolation and preparation of derivatives of 3,5-dimethyl phenol, eugenol, and isoeugenol. These latter represent only a small fraction of the compounds present in this complex mixture.

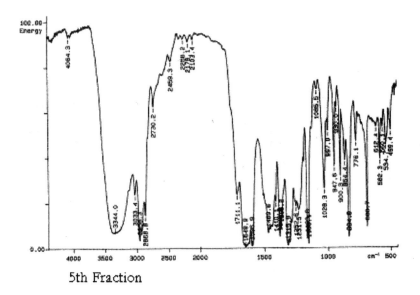

5th Fraction

The final fraction (number 6) was the residue remaining in the stillpot. This was a thick, blood red oil with a strong aromatic odor, representing 3.5% of the original volume distilled.

The strong presence of phenols is indicated by the infrared scan shown below.

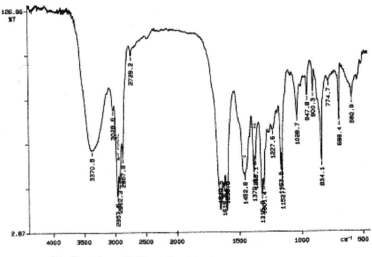

6th Fraction (Stillpot Residue)

The distillate which collected in the cold trap during the sodium acetate distillation was found to consist mainly of acetone with a trace of acetaldehyde. There were additional volatile components in vanishingly small amounts difficult to identify.

All of the data presented above provides an expanded view of the types of materials and their complexity as found in the products of metallic acetate dry distillation.

Several other examples of "Metallic Oils" derived from the acetate path are appended here for comparison.

As noted in the text, the alkali metals and alkaline earths tend to produce a larger quantity of oil.

Fractional distillation of the very small amounts of oil derived from the transition metals proved to be quite difficult. Instead, the Menstruum Foetens obtained from these experiments was separated by solvent extraction.

After distilling out the "White Wine" or acetone fraction, the remaining oily residue in each case was extracted with petroleum ether. The remainder, which was insoluble in the ether, was extracted into acetone.

Upon removal of these extraction solvents, one is left with an ether soluble oil and an acetone soluble oil.

The two following chromatograms are representative of the oils obtained from dry distillation of iron acetate.

In this experiment, weathered pyrite (iron sulfide) was extracted with rainwater. The water was filtered and then evaporated to obtain beautiful green crystals of ferrous sulfate, called green vitriol by the ancient artists.

After purification by recrystallization several times, the crystals were dissolved in water, to which was added a solution of sodium carbonate. The resulting precipitate of iron carbonate was washed with water and then dissolved in concentrated vinegar to produce the acetate.

The recrystallized iron acetate was dried and then subjected to dry distillation.

Approximately 20% of the distillate obtained was acetone, 2.5% were the oils, and the remainder was the watery, acetic "phlegm".

The two solvent extracted oils were analyzed by gas chromatography/ mass spectrometry as shown below.

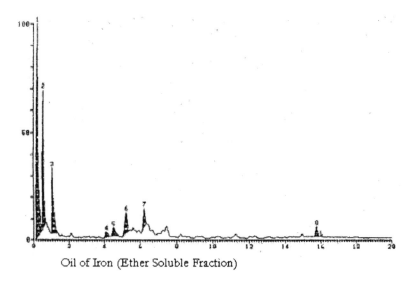

Oil of Iron (Ether Soluble Fraction)

The ether soluble oil contained a mixture of complex ketones and traces of phenolic materials. The two major constituents were identified as 1-methoxy ethanol acetate and 1,1,2,2-tetramethyl cyclopropane.

The acetone soluble fraction proved to be much more concentrated with substituted phenols.

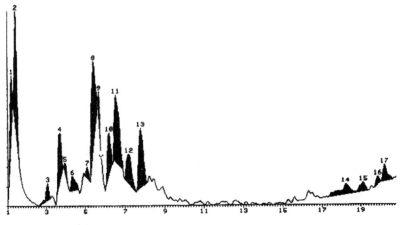

Iron Oil (Acetone Soluble Fraction)

298

Major constituents included 4-methyl-3-pentenoic acid; Phenol; 2-methyl phenol; 2,3-dimethyl phenol; 3,5-dimethyl phenol and 2,4,6-trimethyl phenol.

The next series of scans presents a look at the products from lead acetate dry distillation. This is traditionally an important subject matter, which leads to a Wet Way of producing the Philosopher's Stone.

In this experiment, Galena, the sulfide ore of lead, was roasted to remove sulfur and then extracted with a concentrated vinegar. Slow evaporation of the vinegar produced long, clear crystals of lead acetate, also known as "Sugar of Lead".

A time VS temperature plot is shown below.

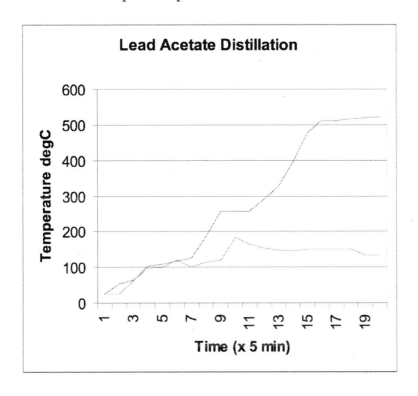

Similar to the sodium acetate distillation, the lead acetate became liquified as the "phlegm" distilled over. At about 30 minutes, the material resolidified and heavy white vapors began filling the apparatus, followed by the appearance of a golden distillate. Towards the end of the distillation, small drops of red oil began coming over.

The distillation of 400 grams of lead acetate, produced 100ml of deep golden liquid in the first receiver and 25ml of clear liquid in the cold trap. These numbers were fairly consistent in other experiments using the same equipment set-up. They are only guidelines, because the set-up itself can alter the results toward greater or lesser yields.

Most of the cold trap liquid proved to be acetone with traces of other unidentified volatiles.

Distillation of the first receiver liquid yielded another 22% acetone. 5% of the remaining liquid was soluble in ether as a golden oil; its chromatogram is shown below.

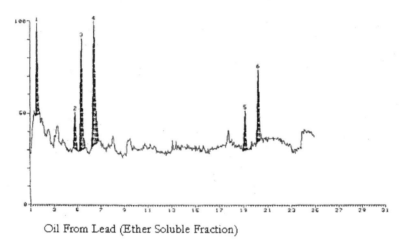

Oil From Lead (Ether Soluble Fraction)

Three of the major constituents of this oil were identified as: 2,4-dimethyl-2-pentanol; 3,5-dimethyl-phenol; and 2,6-dimethyl-2,5-heptadien-4-one.

About 67% of the distillate was a dilute aqueous acetic acid, labeled as the "phlegm". This left about 6% of a deep golden oil which was soluble in acetone. The chromatogram shown below, revealed the presence of 3,5,5-trimethyl-2-cyclohexen-1-one; more 3,5-dimethyl phenol and 5,5-diethoxy-3-pentyn-2-one.

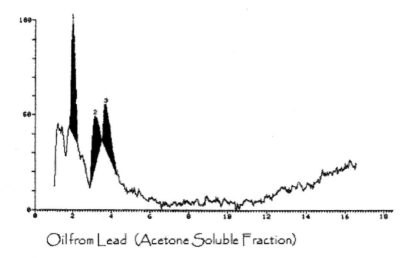

Oil from Lead (Acetone Soluble Fraction)

There was some concern that metallic lead might be present in free form or as an organic compound. Analysis of a sample of the whole distillate by inductively coupled argon plasma optical emission spectroscopy did not reveal its presence, down to a detection level of 40 parts per billion.

As a final example of the acetate path, scans of the oils derived from zinc acetate are included below.

The source material was the sulfide ore of zinc, known as sphalerite. This was roasted and then extracted with concentrated vinegar to obtain the acetate, which was subsequently purified by recrystallization.

During the distillation, a large portion of the matter sublimates as zinc oxide even as far as the first receiver.

Zinc also produces an abundance of volatile spirit (acetone) which is very active towards the extraction of other minerals and metals.

Major constituents of the golden yellow, ether-soluble oil include: 3,3-dimethyl-pentane; 3-hydroxy-butanoic acid; 2,4-dimethyl-2-pentene; 3,5-dimethyl-phenol; and 3-ethyl-phenol acetate.

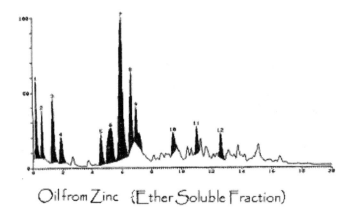

Oil from Zinc (Ether Soluble Fraction)

The acetone soluble oil consisted mainly of the ethyl ester of 4-oxo-2-pentynoic acid; 3,5-dimethyl phenol; and trace amounts of highly substituted compounds, difficult to identify clearly.

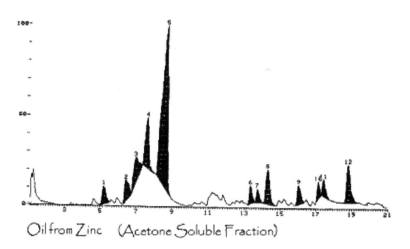

Oil from Zinc (Acetone Soluble Fraction)

As a concluding remark to this section, it will be noticed that many of the compounds formed during the dry distillation are known as self-condensation products of acetone. Acetone can be catalysed in acidic or alkaline environments to link itself together, forming larger molecules often with an oily consistency.

Under alkaline conditions, acetone will undergo reactions to give 4-hydroxy-4-methyl-2-pentanone, commonly known as acetone alcohol.

Under acidic conditions, acetone will link together to form the more stable 4-methyl-3-pentene-2-one, commonly called mesityl oxide. Further reactions can produce phorone and its cyclic homologue, isophorone.

With concentrated acids, acetone can actually trimerize to form the aromatic compound 1,3,5-trimethyl benzene, commonly known as mesitylene.

The catalytic effects are dependent on the particular metal as well as the design of the distillation train and heating rates. Thus each metal plays its part in forming a unique blend of compounds found in the oily distillation product.

APPENDIX II B

Fixed Tincture From Antimony

The following scans illustrate the preparation of a Fixed Red Oil from antimony.

The source mineral was a type of antimony oxide called senarmontite or "Flowers of Antimony". This was produced at the mine, where the antimony ore was roasted in large kilns and the "flowers" sublimated into chimneys, where it was later collected.

Major contaminants in the material included sodium and the other volatile metals arsenic, lead, mercury and zinc.

A long slow calcination up to 450 deg C, followed by washing in water, removed most of the contaminants.

The dried and finely powdered antimony oxide was placed into a glass container and digested with 70% acetic acid at 45 deg C for sixty days. This produced an amber colored liquid extract which was filtered and evaporated into a dark amber gummy resin.

An infrared scan of the resin is shown below.

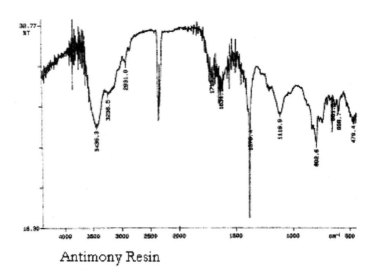

Antimony Resin

The resin was washed with water, dried and separated into two portions. One portion was extracted with ether and one with acetone. Both extractions produced a deep red amber colored oil after removal of the solvent.

Infrared scans indicate that they are essentially the same product, as shown by comparison below.

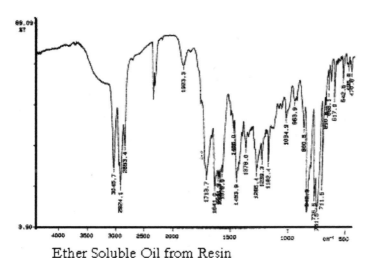

Ether Soluble Oil from Resin

Acetone Soluble Oil from Resin

The oil was dissolved into absolute alcohol and then filtered to produce a Fixed Tincture of Antimony.

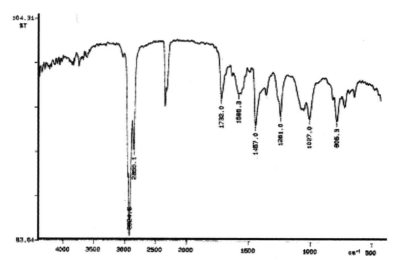

Final Fixed Tincture from Antimony

Analysis for metals content in the final tincture, by plasma emission spectroscopy (ICP), indicated that there were traces of magnesium, sodium and zinc. Antimony and arsenic were below the detection limits of 60 and 30 parts per billion respectively. Later analysis for arsenic by graphite furnace atomic absorption spectroscopy revealed arsenic at 7 parts per billion.

The final tincture was also examined by GC/MS and revealed the presence of a complex mixture of organic compounds. These included residual acetic acid, 1,1-diethoxy-ethane; 4-methyl-3-penten-2-one; ethyl oleate; and 2,6,10,14-tetramethyl-pentadecane as major components.

The chromatogram is shown below.

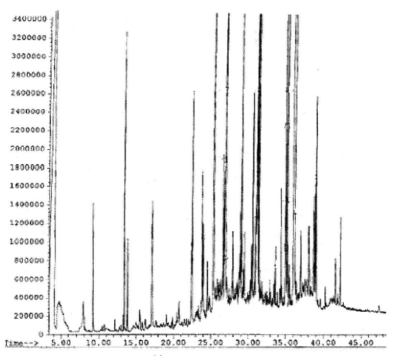

Fxed Tincture from Antimony

APPENDIX II C

Unfixed Oil from Antimony

This next set of scans present the preparation of an unfixed oil from antimony. The source mineral was the same, "Flowers of Antimony", used in the fixed tincture for comparison.

The dried and powdered mineral was reflux extracted with acetone for 9 days. This produced a dark amber liquid extract, which was subsequently filtered off.

Evaporation of the acetone left an amber red oil representing 0.5% of the original weight of antimony extracted.

An infrared scan of this oil is shown below.

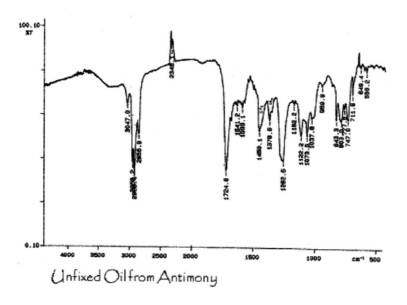

Unfixed Oil from Antimony

Analysis of the oil by gas chromatography reveals its complex composition, as shown below.

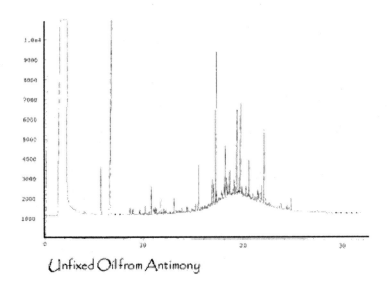

Unfixed Oil from Antimony

On standing, the oil separated into two fractions. On the top was a light golden oil, and at the bottom was a deep red oil.

By volume, there was one part light oil to three parts red oil. The two fractions were separated for individual analysis.

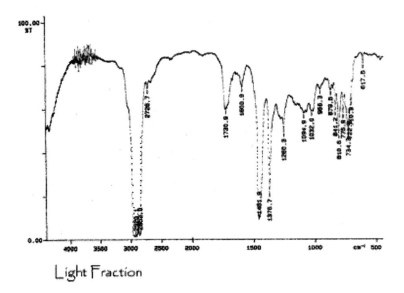

Light Fraction

An infrared scan of the light fraction is shown above, and gas chromatogram is shown below.

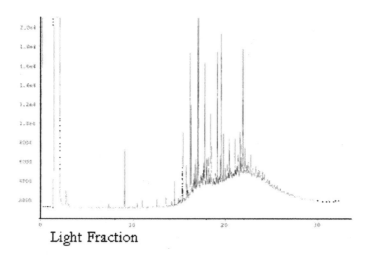

Light Fraction

Analysis of the light fraction by mass spectrometry revealed the presence of 3-ethyl-2-methyl-3-pentanol and 4-hydroxy-4-methyl-2-pentanone as major components.

Most of the trace organics were poorly resolved, so their identification remains tentative at best.

The heavy oil fraction was also examined by infrared and GC/MS. Their scans are shown below.

Mass spectral data indicate the presence of 3,4-dimethyl-2-pentene; 1,1-diethoxy-ethane; 2,5-dihydro-2,5-dimethyl-furan; 2-hydroxy-4-methoxy-benzaldehyde; and more 4-hydroxy-4-methyl-2-pentanone.

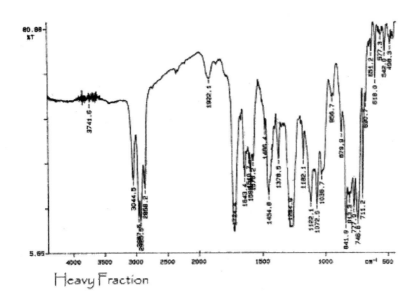

Heavy Fraction

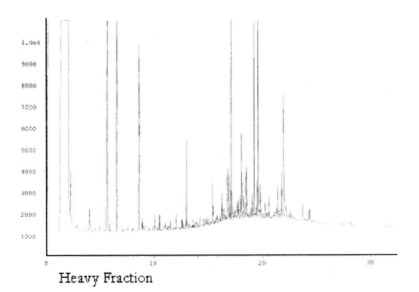

Heavy Fraction

APPENDIX II D

Oil from Gold

This set of scans examines the preparation of an oil from gold as outlined in the text.

The source metal was placer gold flakes obtained by panning river sands. The gold was dissolved in aqua regia, filtered and recrystallized. The resulting brilliant orange crystals were dissolved in water, then neutralized with a solution of potassium carbonate (oil of tartar per deliquiem).

The resulting brown precipitate was then washed with water and dissolved in a minimum of concentrated hydrochloric acid.

This acid solution was concentrated and allowed to crystallize, forming bright golden orange crystals of tetrachloroauric acid, also called gold chloride.

The gold crystals were placed into a glass flask and then covered, with 20 times their weight, with the rectified acetone fraction obtained by dry distillation of lead acetate.

There was a rapid effervesence and the solution became dark amber red in color. After 4 months of digestion, the solution was deep red and oily with some metallic gold settled out.

An infrared scan of the oily liquid is shown below.

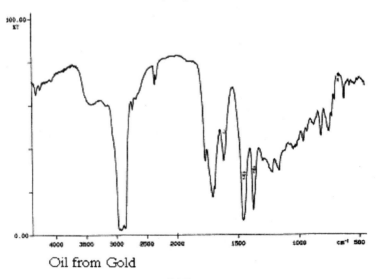

Oil from Gold

Only about half of the gold came out of solution, and this seemed to be the case with other trials using various proportions of gold and acetone. The acetone from sodium and zinc also worked well in the extraction, whereas a commercial acetone worked very slowly and without the initial effervesence.

Examination of the oil was performed by GC/MS as presented in the scan below.

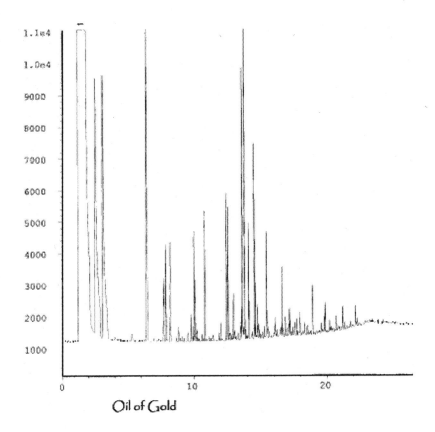

Oil of Gold

A major constituent of the oil was found to be 4-methyl-3-penten-2-one, which is commonly known as mesityl oxide, a condensation product of acetone. In addition, 4-hydroxy-4-methyl-2-pentanone; 1,2,3-trimethyl-benzene; 2,6-dimethyl-2,5-heptadien-4-one; and 2-ethyl-1-pentanol were identified.

313

On standing, the oil separated into two fractions, similar to the unfixed oil from antimony. One was light, at the top, and the other deep red, at the bottom. Their infrared scans and gas chromatograms are presented below for comparison.

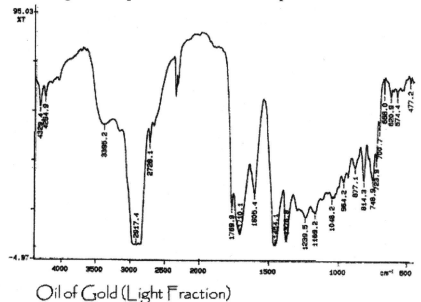

Oil of Gold (Light Fraction)

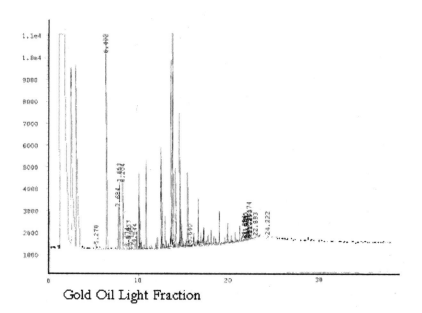

Gold Oil Light Fraction

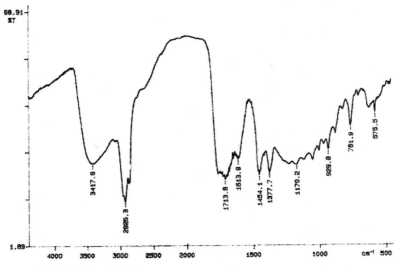

Oil of Gold (Heavy Fraction)

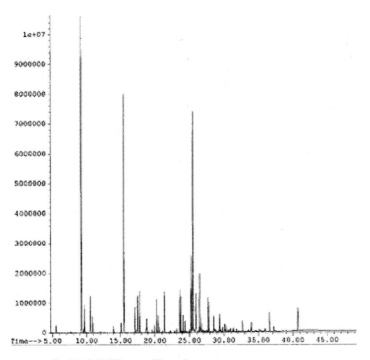

Gold Oil Heavy Fraction

BIBLIOGRAPHY

Allen, Paul M., ed. "Secret Symbols of the Rosicrucians of the 16th and 17th Centuries" in *A Christian Rosenkreutz Anthology*. New York: Rudolf Steiner Publications, 1968.

Anonymous. *Collectanea Chemica*, from the 1893 James Elliot & Co. First Edition. London: Vincent Stuart Publishers LTD, 1963.

Anonymous. *Praxis Spagyrica Philosophica*, English translation from the original 1711 German edition with commentary by Frater Albertus. Salt Lake City, Utah: Paracelsus Research Society, 1966.

Altus. *Mutus Liber*, first published at La Rochelle in 1677. French edition with commentary by Magophon, translated by Kjell Hellesoe. Stavenger, 1985.

Bardon, Franz. *Initiation Into Hermetics*. Dieter Ruggeberg, Wupertal. W. Germany, 1971.

Bartlett, Robert Allen. *Real Alchemy*. Lulu Press, Minnesota, 2007.

Becker, Christian August. *Das Acetone*. 1867. Translated by Shuck and Nintzel, Richardson, Texas: Restorers of Alchemical Manuscripts, 1981.

Dash, Bhagwan. *Alchemy and Metallic Medicines in Ayurveda*. Concept Publishing Co., New Delhi, India, 1986.

Dash, Bhagwan and Lalitesh Kashyap. *Iatro-Chemistry of Ayurveda (Rasa Shastra)*. Concept Publishing Co., New Delhi, India, 1994.

De Lintaut, Henri. *Friend of the Dawn*. 1700 edition translated by Wilson Wheatcroft, India. Richardson, Texas: Restorers of Alchemical Manuscripts, 1982.

Della Porta, Jean Baptiste. *Natural Magic, Book Ten*. New York: 1585. Basic Books, 1957.

Dobbs, B.J.T., *The Foundations of Newton's Alchemy*. Cambridge: Cambridge University Press, 1975.

Dole, Vilas and Prakash Paranjpe. *A Textbook of Rasashastra*. Chaukhamba Sanskrit Pratishthan, Delhi, India, 2004.

Dubuis, Jean. PON Seminars 1992. Translated by Patrice Maleze. Winfield, Illinois: The Philosophers of Nature, 1992.

Essentia: Journal of Evolutionary Thought in Action. Salt Lake City, Utah: Quarterly publication of Paracelsus College, 1980.

Flamel, Nicholas. *Hieroglyphical Figures*. Translated by Eirenaeus Orandus for Thomas Walsley, at the Eagle and Child in Britans Burusse. London, 1624.

Frater Albertus. *Alchemist's Handbook*. Salt Lake City, Utah: Paracelsus Research Society, 1974.

French, John. *The Art of Distillation*. Original 1651 edition published in London. San Francisco, CA: Para Publishers, 1978.

Frawley, David and Vasant Lad. *The Yoga of Herbs*. Lotus Press, Twin Lakes, Wisconsin, 2001.

Fulcanelli. *The Dwellings of The Philosophers*. Translated by Donvez and Perrin. Boulder, CO: Archive Press and Communications, 1999.

Glauber, Johann Rudolf. *The Complete Works of Rudolf Glauber*. Translated by Christopher Packe. Boulder, CO, 1983.

Glaser, Christopher. *The Complete Chymist.* 1677. Richardson, Texas: Restorers of Alchemical Manuscripts, 1983.

Hartmann, Franz. *The Life and Doctrines of Paracelsus.* Los Angeles, California: Mokulume Hill Press, 1972.

Hauck, Dennis William. *The Emerald Tablet: Alchemy for Personal Transformation.* New York: Penguin, 1999.

Holland, Isaac. *A Compendium of Writings* by Johan Isaaci Hollandus. Translated from German by RAMS. Richardson, Texas: Restorers of Alchemical Manuscripts, 1981.

Hurley, Phillip. *Herbal Alchemy.* Lotus Publications, 1977.

Junius, Manfred M. *The Practical Handbook of Plant Alchemy.* New York: Inner Traditions, 1985.

Kervran, Louis. *Biological Transmutations.* London: Crosby Lockwood, 1972.

Kirschweger, Anton. *The Golden Chain of Homer.* Translated from the German 1723 edition by Sigmund Bacstrom in 1797. (Xerox of Bacstrom's handwritten copy.)

Lad, Vasant. *Textbook of Ayurveda.* The Ayurvedic Press, Albequerque, New Mexico 2002

Murien, Petri. *Alchemically Purified and Solidified Mercury.* Translated by Brigitte Donvez. La-Screen. Rasa Vidya Marg, India: Chinchwad, 1992.

Newton, Isaac. *Keynes Ms 64,* King's College Library, University of Cambridge.

Paracelsus. *The Hermetic and Alchemical Writings of Paracelsus,* edited by Arthur Edward Waite. Reprint of the 1894 edition published by James Elliot & Co., London. University Books Inc., 1967.

Paracelsus. *Volumen Medicinae Paramirum.* Translated from the original German by Kurt F. Leidecker. Baltimore: The Johns Hopkins Press, 1949.

Philalethes, Eirenaeus. *An Open Entrance to the Closed Palace of the King.* Richardson, Texas: Restorers of Alchemical Manuscripts, 1981.

Philalethes, Eirenaeus. *The Marrow of Alchemy.* Printed by A.M. for Edw. Brewster at the signe of the Crane in Pauls Churchyard, London, 1654.

Regardie, Israel. *The Philosopher's Stone,* 2nd Edition. St. Paul, Minnesota: Llewellyn Publications, 1970.

Ripley, George. *Liber Secretisimuss.* Richardson, Texas: Restorers of Alchemical Manuscripts, 1982.

Sendivogius, Michael. *The New Chemical Light.* First published in 1608. Richardson, Texas: Restorers of Alchemical Manuscripts, 1982.

The Hermetical Triumph. Printed for Thomas Harris, at the Looking-Glas and Bible, on London Bridge. 1723.

Three Initiates. Kybalion. Chicago: The Yogi Publication Society, 1922.

Trismegistus, Hermes. *Golden Tractate of Hermes.* Salt Lake City, Utah: Para Publishing Co. Inc., 1973.

Valentine, Basil. *Triumphant Chariot of Antimony* with annotations of Theodore Kirkringius MD. Printed for Dorman Newman at the Kings Arms in the Poultry, 1678. Photocopy of original edition.

Valentine, Basil. *The Last Will and Testament of Basil Valentine.* Printed by S. G. and B. G. for Edward Brewster, and to be sold at the Sign of the Crane in St. Paul's Churchyard, 1672.

Weidenfeld, Johannes Segerus. *Secrets of the Adepts.* 1685. Richardson, Texas: Restorers of Alchemical Manuscripts, 1982.

IBIS PRESS
Titles of Related Interest

Real Alchemy
A Primer of Practical Alchemy

Within these pages, we're going to explore alchemy—the "Royal Alchemy." This means we will be exploring Practical Laboratory Alchemy. We will include here the history, theory, and simple practices that anyone can use to prepare herbal and mineral extracts in the ancient tradition. —Robert Allen Bartlett, the author

The book before you is an amazing accomplishment. My friend and fellow alchemist Robert Bartlett has laid bare the secret processes and experiments of our discipline with exceptional clarity and openness. He has exposed the Hermetic origins of alchemy and shown how modern alchemists approach the ancient art. But first and foremost, his book is a revelation of the genuine craft of alchemy as it was meant to be practiced. —Dennis William Hauck, author of *The Sorcerer's Stone*

Real Alchemy by Robert Allen Bartlett is a book for which many have waited a long time. Clean, clear, simple, and easy to read, it provides a foundation in the theory and method of producing plant products ... —Mark Stavish, The Institute for Hermetic Studies, author of *The Path of Alchemy*

An unpretentious masterpiece of anti-obfuscation. *Real Alchemy* is just what the title promises. It is the only book I have ever read on the subject that translates this sacred art and science from metaphor to in-your-face, hands-on objective reality. —Lon Milo DuQuette, author of *The Magick of Aleister*

ISBN: 978-0-89254-150-8
192 Pages, 6" x 9". Paperback $18.95.

The Clavis or Key to the Magic of Solomon
From an Original Talismanic Grimoire
In Full Color
by Ebenezer Sibley and Frederick Hockley
With Extensive Commentary by Joseph Peterson

The Clavis or Key to the Magic of Solomon is a notebook from the estate of Ebenezer Sibley, transcribed under the direction of Frederic Hockley (1808-1885). Sibley was a prominent physician and influential author, who complemented his scientific studies with writings on the "deeper truths." Both Sibley and Hockley were major inspirations in the occult revival of the past two centuries, influencing A.E. Waite, S.L. Mathers, Aleister Crowley, as well as the Golden Dawn, Rosicrucian, and Masonic movements.

This collection reflects Sibley's teachings on the practical use of celestial influences and harmonies. It contains clear and systematic instructions for constructing magical tools and pentacles. *The Mysterious Ring* gives directions for preparing magic rings. *Experiments of the Spirits Birto, Vassago, Agares,* and *Bealpharos,* show how to call upon angels and spirits, and perform crystal scrying. *The Wheel of Wisdom* gives concise directions for using celestial harmonies. The final text, the *Complete Book of Magic Science,* is closely akin to the *Secret Grimoire of Turiel,* but more complete.

The manuscript reproduced here is the most accurate and complete known, very beautifully and carefully written. With extraordinary hand-colored seals and colored handwritten text. (Blank pages have been eliminated so that the 384 page original has been reproduced here as 288 pages.)

ISBN: 978-0-89254-159-1
400 Pages, 7.5" x 9". Hardcover $95.00.

Arbatel
Concerning the Magic of the Ancients
Newly translated, edited and annotated
by Joseph H. Peterson

Arbatel is one of the most influential magical texts. Its many aphorisms are designed to guide us through a transition from an ordinary life to a magical life. It teaches that God created angels to help people, but that we need to learn how to attract and call upon them for help, both spiritual and material. The angels not only can help, protect, and heal, but the higher sciences can only be learned directly from them.

Arbatel also insists on the need to avoid superstition, and being constantly tricked and manipulated by evil forces which are always working against us.

This new edition includes the first English translation published since Turner's in 1655, and a fresh analysis utilizing important new research by Carlos Gilly, Antoine Faivre, and others. It illuminates many obscure points in the text, and explains the magical techniques employed, and its influence on esoteric literature, including the grimoires and the Theosophical movement. Includes illustrations, bibliography, index, and original Latin text.

ISBN: 978-0-89254-152-2
128 Pages, 6" x 9". Hardcover $35.00.